CONTENTS

ACKNOWLEDGMENTS

Historians have long heeded the words of the French scholar Charles Seignobos, who said, "History is made with documents. . . . No documents, no history." I am indebted to organizations whose documents made this book possible, particularly two societies that preserve the St. Louis area's past.

The State Historical Society of Missouri houses the Kay Drey Mallinckrodt Collection, compiled by an antinuclear activist who documented the radioactive contamination of the Coldwater Creek watershed. (Drey made notations on the documents, occasionally printing "Gem" on items she considered noteworthy.) The historical society also maintains the Edward Mallinckrodt Jr. Papers, which detail the life of the industrialist, conservationist, and philanthropist who was board chairman of the Mallinckrodt Chemical Works from 1928 to 1965.

Another organization, the private Missouri Historical Society, introduced me to the early history of north St. Louis County through the papers of Richard Graham. A former soldier and Indian agent, Graham established the Hazelwood plantation in the Coldwater Creek watershed during the early nineteenth century. His farm records illuminate the early environment of the region, as well as the lives of enslaved Black people who were among its earliest nonindigenous residents. The Missouri Historical Society also maintains 1960s-era photographs of the Paddock Hills neighborhood in Florissant, where the Coldwater Creek Facts group originated. Janell Rodden Wright, a founding member, supplemented the archival information by granting me an interview on recent neighborhood developments and important initiatives of her group.

Libraries also made substantial contributions to this book. The staff of the main branch of the St. Louis County Library offered me critical maps and genealogical data on early residents of the Coldwater Creek watershed. Employees of the Lovejoy Library at Southern Illinois University Edwardsville helped me locate secondary sources in the United States and abroad to put the primary documents in context.

Several newspapers and other media outlets were very helpful; however, special credit goes to the *St. Louis Post-Dispatch* and reporters Carolyn Bower, Louis J. Rose, and Theresa Tighe for their 1989 investigative series on nuclear contamination in the St. Louis area. Finally, I was assisted by documents from several government offices or agencies. They include the U.S. Army Corps of Engineers, U.S. Department of Energy, U.S. Department of Health and Human Services, Formerly Utilized Sites Remedial Action Program, Missouri Department of Natural Resources, National Park Service, St. Louis County Department of Health, St. Louis County Department of Planning, and Metropolitan Sewer District of St. Louis. The documents include a number of reports that were critical to my work. They are:

Coldwater Creek, Missouri: Feasibility Report and Environmental Impact Statement, 1986;

Faisal Khan, "Health Advisory: Report of Coldwater Creek Community Exposures Released," 2018;

Missouri Department of Natural Resources Water Protection Program, Bacteria Total Maximum Daily Load (TMDL) for Coldwater Creek, St. Louis County, MO, 2014;

Record of Decision for the North St. Louis County Sites, 2005;

Shumei Yun et al., *Analysis of Cancer Incidence Data in Coldwater Creek Area, Missouri, 1996–2004*, 2013;

St. Louis Site Remediation Task Force Report, 1996;

U.S. Department of Health and Human Services Public Health Service Agency for Toxic Substances and Disease Registry, Division of Community Health Investigations, *Evaluation of Community Exposures Related to Coldwater Creek St. Louis Airport / Hazelwood Interim Storage Site, North St. Louis County, Missouri*, 2019.

As my manuscript was assessed, revised, and readied for publication by the University of Georgia Press, I received excellent support from Mick Gusinde-Duffy, executive editor for scholarly and digital publications. I also benefited from the helpful suggestions of scholars who gave my work a blind review.

Two people generously gave of their time to read my drafts and offer suggestions. My heartfelt thanks go to Laurel Puchner, professor in the Department of Educational Leadership at Southern Illinois University Edwardsville; and my husband, Jim Morice, who came of age in the Coldwater Creek watershed and later covered the St. Louis metropolitan area as a journalist. My brother-in-law Don Morice also provided valuable assistance with the photo images in this book. Longtime friend Kathryn Davis offered technical support. Friends and former St. Louis journalists Andrew Wilson and Beth Ann Wilson provided helpful commentary as the project neared production.

Finally, I am indebted to people who shared their memories of growing up near Coldwater Creek as I wrote this book. They include my brother Jim Cunningham, my cousin Russell Viehmann, my former neighbor Linda Voss Keller, and my longtime friends Judy Weaver Failoni and Kathy Loberg Riley. I hope the contributions of these organizations and individuals provide a clearer understanding of what happened at Coldwater Creek and increase our resolve to be better environmental stewards in the future.

NUKED

INTRODUCTION

This book began with a life-changing event that only revealed its significance decades later. In September 1957—at the age of nine—I traveled with my parents and two brothers along Route 66 in our family car, a Hudson Super Jet. The two lanes of concrete and asphalt had been laid in the 1920s to connect the American heartland to the West Coast. However, by 1957 Route 66 had become a cultural icon that offered (according to boosters) adventure and a path to the promised land. In keeping with that outlook, our family traveled the highway with optimism, anticipating a new life in Missouri, where my father, an aeronautical engineer, had accepted a job. Years later I would learn that our confidence in the future was somewhat misplaced.

Just three days prior, we had vacated the only home I had ever known, in the city of Detroit. After spending the first night in Toledo, we traversed Ohio, Indiana, and Illinois before crossing the wide Mississippi River into Missouri. Our Super Jet crawled across the narrow Chain of Rocks Bridge as traffic slowed to accommodate a unique feature in its steel-truss design: the bridge did not follow a straight line, but forced motorists into a thirty-degree curve midway in the span. During the long crossing, I wondered what other surprises might await us in our new home in a community called Florissant. Founded in 1787 as a French colonial village, Florissant was fourteen miles northwest of St. Louis. Although the buildings and grid of the "Old Town" were still intact in 1957, the once-sleepy community was quickly being surrounded by new suburban homes.

My family typified Florissant's new residents, many of whom worked in the defense industry at or near Lambert Field, the location of St. Louis's municipal airport. Among our neighbors were military officers, navy test pilots, and engineers who worked at McDonnell Aircraft (my father's new employer). Many hailed from other parts of the United States. At a time when most U.S. workers expected to remain with one employer until retirement, these newcomers were prepared to move again. They understood their career trajectory would be determined largely by the government's choice of companies for military contracts. Accordingly, residents wanted newly constructed single-family homes that could be sold quickly if needed—as well as central air-conditioning to mitigate St. Louis's hot, humid summers.

A Postwar Economy

My family's experience—and that of our neighbors—reflected a major reordering of the U.S. economy after World War II. Many companies transitioned from wartime production to meet pent-up consumer demands, while others continued to manufacture materiel for the Cold War.

My father's previous employer, the Hudson Motor Car Company of Detroit, had focused on war production while the global conflict raged. The company continued to receive aircraft contracts in peacetime but eventually returned to producing automobiles. In 1954 Hudson merged with Nash-Kelvinator to form American Motors Corporation (AMC) in what was then the biggest corporate merger in U.S. history. AMC tied its fortune to automobile production (eventually getting out of the airplane business), while McDonnell Aircraft, at St. Louis's Lambert Field, continued to design and build fighter jets for the military. After McDonnell hired my father, the company's housing director showed my mother a variety of properties in north St. Louis County. My parents chose the Paddock Hills neighborhood of Florissant, where they rented a house for nine months while building their own home.

Upon arrival in Florissant, each member of our family found a means of connecting with the new community. My father settled into his job and eventually pursued a photography hobby through the McDonnell Camera Club. My mother immediately became involved in passing a school referendum. In less than five years, she was elected to the Ferguson-Florissant School District Board of Education—a post she held for fifteen years. My brother John (age twelve) won a prize for his Halloween painting on the window of Florissant's only shoe store. I took dance lessons near a working blacksmith shop on Florissant's main street. My brother Jimmy (age five) bonded with a neighborhood dog. We continued to explore our new and ever-changing environment that featured suburban housing, bucolic farmland, historic structures, and construction sites. We did not expect that it would be our parents' last home.

A Secret

My family moved into our new house in June 1958. It was situated in the Coldwater Creek watershed, on a gentle slope approximately one thousand feet from a tributary of the main channel. Unbeknownst to us, the creek had flooded extensively one year earlier, and it would continue to flood to a somewhat *lesser* extent thereafter. In each case the floodwaters backed up in the tributaries as they moved sediment downstream. Years later I would learn that both the water and sediment were radioactive—contaminated by uranium processing wastes from the atomic bomb dropped on Hiroshima. Under the top-secret Manhattan Project, Mallinckrodt Chemical Works conducted the refining in St. Louis, a long way from our home. However, the U.S. government "stored" the radioactive wastes on a 21.7-acre property called the St. Louis Airport Storage Site beginning in 1946. As the name suggests, the land was next to the municipal airport and adjacent to Coldwater Creek, some nine miles upstream from where we lived. The contaminants leached into the water and soil, exposing the downstream population to ionizing radiation.[1] The result was a health crisis leading to illness and death for many north St. Louis County residents. Some of my immediate family members were among the lives lost.

This book explores a simple question concerning the radiological exposure at Coldwater Creek: *How did it happen?*

Overview

Nuked: Echoes of the Hiroshima Bomb in St. Louis examines a major public health crisis in north St. Louis County, Missouri, resulting from U.S. nuclear weapons production in World War II. The book explores the interaction of human and natural forces that created the radiological contamination of land, water, and air in the Coldwater Creek watershed and led to widespread illnesses and deaths.

The discussion focuses on the Manhattan Project, a top-secret initiative of the U.S. government to produce an atomic bomb to end World War II. The Mallinckrodt Chemical Works of St. Louis played a key role in that effort by refining uranium for the world's first controlled, self-sustaining nuclear chain reaction—and for the Hiroshima bomb. Since uranium refining produced enormous amounts of waste, government officials looked for a place to "store" the byproducts. They chose a 21.7-acre property in north St. Louis County in what was then a rural area. The land was located next to the municipal airport and bounded by Coldwater Creek. For ten years, truckers transported 55,000 drums filled with 30 or 50 gallons of the radioactive material, then arranged them end to end over most of the storage site. The truckers also deposited vast quantities of uranium processing waste in bulk—on the ground, uncovered, and next to the creek. Wind, rain, and snow eroded the piles, causing the contaminated material to leach into the waterway. The drums also rusted and leaked, adding their contents to the mix. As a result, radionuclides entered Coldwater Creek, polluting the water and sediment during its 13.7-mile journey to the mouth of the Missouri River. For decades, residents of the creek's 47-square-mile watershed remained unaware of their exposure.

Decision-makers for the Manhattan Engineer District (Manhattan Project) and its successor agencies were responsible for addressing the environmental and health crisis at Coldwater Creek. However, these officials' actions often contradicted their espoused beliefs. They valued democracy but made top-down decisions with little congressional oversight. They championed a free press while lying to journalists and therefore to the public. They valued secrecy while storing radioactive materials in the open, in plain sight. They sent troops overseas to protect the country but exposed the homeland to deadly toxins. Through it all, the officials justified their actions as necessary for U.S. victories in World War II and the Cold War. Unfortunately, subsequent generations are still paying for the mistakes of the nation's nuclear weapons program during these years.

Thesis and Influences

The book's thesis attributes the crisis at Coldwater Creek to the Manhattan Project's prioritization of politics and expediency over health and the environment. It contends that winning the war and producing vast quantities of nuclear weapons were more important to the Man-

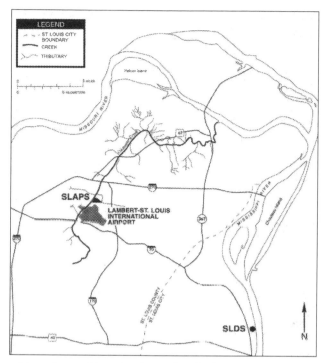

Note the creek's main channel and tributaries and their
proximity to the St. Louis Airport Storage Site (SLAPS).
Mallinckrodt Chemical Works refined uranium at the
St. Louis Downtown Site (SLDS) before sending the
radioactive processing wastes to SLAPS.
U.S. ARMY CORPS OF ENGINEERS

hattan Project and its successors than safeguarding U.S. citizens and
the biosphere in which they lived.

In addition to the sources discussed in the acknowledgments, two
works inform the book's theoretical framework. They are *The Search
for the Ultimate Sink: Urban Pollution in Historical Perspective* by
Joel A. Tarr,[2] and *Plutopia: Nuclear Families, Atomic Cities, and the
Great Soviet and American Plutonium Disasters* by Kate Brown.[3] Tarr
focuses on the impact of urban industrial wastes and contends they
became a major concern only after World War II. In earlier times,
these wastes were viewed as existing *inside* the workplace. However,
in the postwar era—and extending through the passage of environmental legislation in the 1970s—professionals in the United States became increasingly concerned with the industrial pollution of air and

water. When such contamination was no longer an acceptable option for waste disposal, industries turned to polluting the land and eventually replaced the open dump with the sanitary landfill.

Tarr's discussion of industrial waste provides a context for understanding the contamination of Coldwater Creek. Uranium byproducts from the first atomic bomb remained inside the Mallinckrodt factory until 1946, when they were moved for "storage" on rural land adjoining the St. Louis airport and abutting the creek. This odd arrangement resulted from U.S. negotiations with a Belgian company that owned a Congolese mine with the world's purest uranium. The United States agreed to eventually return the derivatives of uranium refining so that the company could process them for other elements. However, once on the land, the radioactive byproducts polluted large swaths of north St. Louis County. Eventually, the contaminants were dispersed to other locations in Missouri and beyond as people continued to search for a new place of deposit—or, in Tarr's words, a "sink."[4]

Kate Brown's *Plutopia* connects two model communities that produced plutonium for weapons manufacture during the Cold War: Richland, in eastern Washington State near the Hanford nuclear site, and Ozersk, in the southern Russian Urals. Plant managers at both locations pushed to produce as much plutonium as possible, thereby contaminating the surrounding landscape. In the discussion of Richland, Brown expands on Tarr's book by contending that scientists at the Hanford Site wanted to turn the *entire Columbia River Basin* into a sink for depositing radioactive toxic waste. They believed the contaminants would scatter into the air, be diluted by the water, and vanish into the soil, becoming harmless.[5]

Although no one intended to turn the entire Coldwater Creek watershed into a vast sink, officials voiced the same arguments as they minimized the impact of radionuclides at that Missouri location. The U.S. Department of Energy was aware in the mid- to late 1970s that radioactive material was entering the creek yet wrongly believed the water would dilute it and make it indistinguishable during seasonal floods.

Organization

Nuked is organized into three parts, each with two chapters. Part 1, "The Contamination," discusses the military threat that persuaded the United States to create the Manhattan Project. This section explores Mallinckrodt's role in that initiative, as well as the U.S. agreement to import uranium ore from the Belgian Congo. Part 2, "The Dissemination," explains the "storage" of uranium byproducts near the St. Louis airport and the resulting contamination of Coldwater Creek. This development was made worse by postwar construction practices that plowed up tainted soil and spread it in new suburban neighborhoods. Part 2 also discusses the failure of the Manhattan Project's successor agencies to stop the spread of the airport contamination to new sites. Giving false assurances of safety, government officials were challenged by residents who wanted the pollutants somewhere other than their neighborhood. Finally, part 3, "The People," presents the efforts of twenty-first-century advocates who pushed government officials to test for contamination in the northern reaches of the Coldwater Creek watershed. They finally received an admission from the federal government that linked the contamination of the waterway to certain types of cancers. The book concludes by presenting human stories of the Coldwater Creek tragedy and how communities in Kentucky, Colorado, and Oregon (also participants in the U.S. nuclear program) were similarly handled by the Manhattan Project and its successors.

The book also contains three appendixes: an evaluation of community exposures by the Agency for Toxic Substances and Disease Registry; a health advisory from the St. Louis County Department of Public Health; and resident questions posed to the Army Corps of Engineers, the group currently charged with cleaning up Coldwater Creek. A timeline and glossary of terms follow the appendixes.

It is hoped that this book will provide new perspective on civilian casualties of the $25 billion undertaking that was the Manhattan Project.

PART ONE

The Contamination

CHAPTER 1

The Secret Weapon

I told you it [building an atomic bomb] couldn't be done without
turning the whole country into a factory. You have done just that.

—Physicist Niels Bohr to his colleague Edward Teller

St. Louis entered the atomic age in a genteel setting. On April 17,
1942, two longtime friends met for lunch at the downtown Noonday Club, a bastion of the city's power elite. The renowned physicist
Arthur Holly Compton did most of the eating and talking while Edward Mallinckrodt Jr.—a chemist, conservationist, and chairman of
the board of Mallinckrodt Chemical Works—listened and ate his usual
bowl of cold cereal. The meal ended with a handshake that signaled
Mallinckrodt's partnership in creating a secret weapon that would
shake the world.[1]

It was not the first historic meeting in the Noonday Club's forty-
nine-year history. In 1927 several men met there to arrange financial
backing for Charles Lindbergh's transatlantic flight in the *Spirit of St.
Louis*. Before taking off for New York and Paris, Lindbergh thanked
the backers by circling his plane over the building, pointing the nose
down, and flying close to the flagpoles "so everyone there would hear
[his] engine."[2] Compton and Mallinckrodt's exchange at the Noonday
Club was considerably more subdued, being prompted by a national
emergency.

Just four months after the United States' entry into World War II,
Adolf Hitler was beating the Allies in Europe.[3] Japan was on the verge
of driving the United States from the Philippines. Especially concerning were American intelligence reports that German scientists were

two years ahead of the Allies in developing the "ultimate weapon" to ensure Hitler's victory. Speaking on behalf of the federal government, Compton urged Mallinckrodt to embark on an untried project that three companies considered too dangerous. Scientists at the University of Chicago needed forty tons of pure uranium for an experiment to create the first controlled, self-sustaining nuclear chain reaction. No more than one-half cup of uranium existed in the United States at the time. However, it could be extracted from uranium oxide ore using ether—a highly volatile and flammable liquid. If successful, the experiment could lead to America's development of an atomic bomb before Germany and result in an Allied victory in World War II.[4]

As Edward Mallinckrodt Jr. listened to Compton's request, he might well have reflected on the evolution of his family business. In 1831 German immigrant Emil Mallinckrodt had purchased a plot of land north of the city, in an area later annexed by St. Louis. The business applied chemistry to farming and specialized in orchards and vineyards. As the only chemical company west of Philadelphia, Mallinckrodt captured emerging Western markets and by 1890 was a chief producer of anhydrous ammonia, an important chemical used in refrigeration. The establishment manufactured morphine and codeine in 1898 and began making barium sulfate—a key component in contrast media–enhanced images for X-ray diagnoses—in 1913. The following year Mallinckrodt Chemical Works perfected a new method of storing ether, the first modern anesthetic. Over time, the company specialized in ether production for anesthesia and enjoyed a reputation for pure chemicals. Compton had made a smart choice.[5]

Immediately after lunch with his friend, Edward Mallinckrodt Jr. convened his team to implement the top-secret endeavor, which was affiliated with an initiative now commonly called the Manhattan Project.

Following the Science

During the lunch Compton reflected on his own evolving support of the war effort. His father was a Presbyterian minister and philosophy professor. His Mennonite mother, who favored missionary work, became the national Mother of the Year in 1939. Strongly influenced by her pacifism and by isolationist sentiments in the country at large, Compton also recognized his elder brother Karl's scientific prominence in winning a top-secret research laboratory for Massachu-setts Institute of Technology (MIT), where he was president. Arthur Compton had his own impressive resume, which included a PhD from Princeton, the 1927 Nobel Prize for physics, a professorship at Wash-ington University in St. Louis (where he chaired the physics depart-ment), and a professorship in physics at the University of Chicago.[6] Later he would give the following explanation for supporting the U.S. entry into the war as peace prospects dimmed:

> In 1940, my forty-eighth year, I began to feel strongly my responsi-bility as a citizen for taking my proper part in the war that was then about to engulf my country, as it had already engulfed so much of the world. I talked, among others, with my minister in Chicago. He wondered why I was not supporting his appeal to the young peo-ple of our church to take a stand as pacifists. I replied in this man-ner: "As long as I am convinced, as I am, that there are values worth more to me than my own life, I cannot in sincerity argue that it is wrong to run the risk of death or to inflict death if necessary in the defense of those values."[7]

Shortly after Compton arrived at this position, a science adminis-trator named Vannevar Bush recruited him into government service.[8] Compton's subsequent actions, including his recruitment of Mallinck-rodt, were strongly influenced by scientists' repeated attempts to har-vest the enormous energy inside an atom.[9]

As early as 1933, Hungarian physicist Leon Szilard realized the possibility of a nuclear chain reaction to release that energy; yet five years later scientists had not identified the elements that would cre-ate the reaction.[10] Then Otto Hahn and Fritz Strassmann made an unexpected discovery when bombarding elements with neutrons in their Berlin laboratory. The two chemists found that while the nu-clei of most elements changed somewhat during bombardment, ura-nium nuclei changed greatly and broke into two roughly equal pieces. Shortly after that experiment (conducted in 1938), Austrian scientist Lise Meitner and her nephew Otto Frisch advanced the theory that uranium breaks into smaller atoms when bombarded by neutrons. They used the term "fission" to describe the process. Physicists from Britain, France, Italy, Russia, and the United States rushed to dupli-cate Hahn and Strassmann's experiment.[11]

On August 2, 1939, the world-renowned physicist and mathematician Albert Einstein warned President Franklin D. Roosevelt of Germany's efforts in uranium research:

> It may become possible to set up a nuclear chain reaction in a large mass of uranium, by which vast amounts of power and large quantities of new radium-like elements would be generated. Now it appears almost certain that this could be achieved in the immediate future.
>
> This new phenomenon would also lead to the construction of bombs, and it is conceivable—though much less certain—that extremely powerful bombs of a new type may thus be constructed. A single bomb of this type, carried by boat and exploded in a port, might very well destroy the whole port together with some of the surrounding territory. However, such bombs might very well prove to be too heavy for transportation by air.[12]

The German-born scientist recommended that FDR secure "a supply of uranium ore" and prioritize critical research that, at the time, was "being carried on within the limits of the budgets of University laboratories." Einstein added that the U.S. supply of uranium was very poor and available only in "moderate quantities." While "some good ore" existed in Canada as well as "the former Czechoslovakia" (where Germany had taken over the mines), the most important source of uranium was in the Belgian Congo. Einstein was a pacifist, but he wanted to prevent Germany from having sole possession of the new, destructive power of an atomic bomb.[13]

Less than one month after Albert Einstein had written those words, Hitler invaded Poland, and World War II began.

Evolution of the Manhattan Project

FDR responded to Einstein's letter by establishing three separate groups that gradually expanded the U.S. nuclear program. First was the "modest program of basic research" under the Uranium Committee of the National Research Council. Working under the leadership of Lyman J. Briggs, director of the National Bureau of Standards, the group coordinated research into the properties of uranium but spent little money on the effort.[14] In April 1940 its members were startled to learn that the Kaiser Wilhelm Institute of Physics in Berlin had begun an extensive research program on uranium. The information spurred Roosevelt to expand uranium-related activities, and by June the Uranium Committee was reconstituted under the newly created National Defense Research Committee (NDRC). Led by Vannevar Bush, a former vice president of MIT, the NDRC investigated the fissile properties of uranium-235 and uranium-238 and began to evaluate methods for separating isotopes of that element. The NDRC also coordinated research into the fissile properties and potential weapons applications of a newly discovered element that was later called "plutonium."[15]

However, by the summer of 1941, NDRC was absorbed by the Office of Scientific Research and Development (OSRD), reflecting a higher priority on the military application of atomic energy. As a scientific research organization, the NDRC had been unable to "fill the gap" between research and procurement and found itself competing for resources with laboratories operated by the army and navy. The OSRD, by contrast, was located within the Office of Emergency Management in the Executive Office of the President, under Bush's direction. James Bryant Conant, Harvard president, replaced Bush at the NDRC, which continued under the OSRD umbrella. Bush and Conant directed research that continued to explore different isotope separation methods while working on the production of plutonium, a radioactive element formed from uranium-238.[16]

In 1942, OSRD was succeeded by the Manhattan Engineer District (MED). Although OSRD had made progress in separating radium isotopes and producing plutonium, its members lacked expertise in designing and building production-style plants. Kevin O'Neill writes, "Far from pilot facilities . . . these full-scale plants were designed from the start to generate enough materials for a sizable stockpile of atomic bombs." The MED would remain in existence until 1947, when it was replaced by the Atomic Energy Commission (AEC).[17]

Meanwhile, Arthur Holly Compton was busy recruiting fellow scientists for top-secret work. Only two months before his lunch with Mallinckrodt, Compton had begun working at the Metallurgical Laboratory at the University of Chicago. The name disguised the true nature of the work: conducting research for an atomic bomb. Toward that end, Compton identified scientists in the United States and Europe, including some who had fled fascist countries. Among them was Enrico Fermi, who immigrated to America with his Jewish wife, Laura Capon Fermi. At Chicago, he directed an experiment to produce the world's first controlled, self-sustaining nuclear chain reaction, which, if successful, would confirm the possibility of harnessing the atom's energy.[18]

As early as October 9, 1941, President Roosevelt had approved the top-secret development of the atomic bomb. Initial funding for the project was channeled into a "secret dummy account" to avoid detection by a foreign spy and to bypass the legal requirement of obtaining congressional authorization.[19] By June 1942, Roosevelt and top U.S. government officials had approved a plan to assign the Army Corps of Engineers the task of producing an atomic weapon before the war's end. Two months later, the chief of engineers issued a general order establishing "a new engineer district, without territorial limits to be known as the Manhattan District . . . with headquarters at New York, New York, to supervise projects assigned to it by the Chief of Engineers." On September 17, 1942, Colonel (quickly promoted to Brigadier General) Leslie R. Groves was put in charge of the Manhattan District, also known as the Manhattan Engineer District (MED). The Manhattan Project was the top-secret group within the MED that focused on the development of atomic bombs.[20] According to Susan Williams, senior advanced fellow at the University of London, "Secrecy was already the watchword of the whole enterprise and this name was chosen as a way of avoiding giving any clues about its concerns."[21]

Upon his appointment, General Groves began building the core of the U.S. nuclear weapons production complex. He was a straightforward and efficient man who thought large staffs were "conducive to inaction and delay." Only a few people (and in some cases only Groves) made the crucial decisions on atomic installations and their practices during the war years.[22] One subordinate described Groves as having "an ego second to none" and "tireless energy," adding, "He had absolute confidence in his decisions and he was absolutely ruthless in how he approached a problem to get it done. But that was the beauty of working for him—that you never had to worry about the decisions being made or what it meant."[23] Having previously overseen all domestic Army construction (including the building of the Pentagon) during the mobilization for World War II, Groves immediately recruited industrial and construction firms—Union Carbide & Carbon, Tennessee Eastman, DuPont, Standard Oil, M. W. Kellogg, Chrysler, Monsanto, Stone & Webster, J. A. Jones, and others—to design, build, and operate the first-of-a-kind production facilities. Amid scientists' concerns about the military's dominant role in the project (and implications for the ultimate use and control of the weapon), Roosevelt and British prime minister Winston Churchill agreed to keep the atomic bomb a secret.[24]

Congressional leaders supported the need for secrecy and largely gave Groves a blank check for MED expenses. Senior War Department and military officials briefed House leaders in February 1944 (seventeen months after the project had gotten underway) about its "finances, construction, procurement, and schedules" and "overall connection to the war." They avoided sharing too many details. House members gave their approval without reservation and indicated further explanation would be unnecessary. The Senate leadership concurred. As a result, most members of Congress remained in the dark about the MED's activities.[25]

A strict "need-to-know" policy ensured the safeguarding of information in all aspects of the Manhattan Project. Compton reported, "Only about six men in the U.S. Army are permitted to know what is going on, including Secretary of War Stimson!" Since the top scientists were among the few civilians who did know what the project was about, they received bodyguards from a special group of Counter Intelligence Corps agents who were really spying on the individuals they protected.[26]

On the Home Front

When Edward Mallinckrodt Jr. returned from the Noonday Club, he and his team didn't bother with blueprints, preferring to sketch their ideas on paper or chalk them on the floor or wall. Within a day or two, carpenters and pipe fitters were using the ideas of engineers and chemists to build equipment. When engineers determined the project needed stainless steel kettles and motors (neither of which were available in wartime), the company dismantled one of its production lines in New Jersey and shipped it to St. Louis. Employees worked around the clock so that, within a month, Mallinckrodt Chemical Works was producing a ton of pure uranium daily. Although Mallinckrodt and the Manhattan Engineer District eventually had a formal contract, it was not finalized until the company had processed much of the uranium. Because of the project's urgency, the U.S. government placed a low priority on finding ways to treat the radioactive waste it generated. This omission occurred despite indications that the volume of wastes and residues would be enormous.[27]

On December 2, 1942—in a former squash court under the bleachers at the University of Chicago's Stagg Field—Enrico Fermi directed the world's first controlled, self-sustaining nuclear chain reaction.

The uranium in the experiment had been purified at the Mallinck-rodt Chemical Works in St. Louis. Fermi compared an atomic chain reaction to "the burning of a rubbish pile from spontaneous combustion." He explained, "Minute parts of the pile start to burn and in turn ignite other tiny fragments. When sufficient numbers of these fractional parts are heated to the kindling points, the entire heap bursts into flames."[28]

Mallinckrodt had done an extraordinary job of supporting Fermi's work. Beyond the constant risk of explosion (heightened by a lack of refrigeration at the plant and a constant need for ice in the purification process), the biggest problem was obtaining massive volumes of uranium, which largely existed outside the United States.[29]

Shinkolobwe

In its purest form, uranium was found in the Belgian Congo in the mineral pitchblende (now called uraninite). The source was the Shinkolobwe Mine, owned and operated by a Belgian company, the Union Minière du Haut Katanga. Compared to other mines in the world, Shinkolobwe's uranium supply was exceptional. Some of its ore yielded a uranium oxide content of 75 percent (with an overall average in excess of 65 percent). By contrast, the ore mined in South Africa was 0.03 percent uranium oxide—and in Canada or Colorado, a mere 0.02 percent. Large amounts of uranium were required to produce a nuclear reaction and therefore an atomic bomb.[30]

Described as "a huge, open gash about a half-mile square," the mine had "terraced sides that went down 225 feet." It had been used as a source of radium (a byproduct of uranium), but it closed in 1937 when its stockpiled ore seemed adequate to satisfy the anticipated world demand for thirty years. The mine fell into disrepair and became flooded.[31]

Fortunately for the United States, Edgar Sengier, managing director of the Union Minière du Haut Katanga, came to New York from Brussels in the fall of 1939. Late in 1940, when the German seizure of much of Africa seemed likely, Sengier ordered 1,200 tons of high-grade uranium oxide ore to be shipped from Shinkolobwe to New York for safekeeping in a Staten Island warehouse. To handle the shipments, he established African Metals Corporation as Union Minière's commercial subsidiary. Although the United States acquired 100 tons of the ore, Union Minière kept ownership of the radium contained therein.

America had an option to purchase the remaining 1,100 tons of uranium oxide ore on Staten Island and twice that amount from the Belgian Congo stockpile.[32]

Meanwhile, American and British officials tried to persuade the Belgians (whose government was in exile in London) to reopen the Shinkolobwe Mine. On September 26, 1944, the three countries reached a tripartite agreement in which Belgium granted the United States and Britain an option on all its uranium and thorium resources. The arrangement could continue for the period required to carry out ore contracts under the agreement's terms plus ten years. However, the Belgian government reserved the right to retain such ore as needed for its own scientific research, energy, and commercial purposes.[33]

Importantly, the United States purchased only uranium from the pitchblende, although the mineral also contained high amounts of radium and thorium. All radium-containing residues were to be held for eventual return to African Metals Corporation. This meant uranium byproducts were "stored" in America and not disposed of immediately after processing.[34] Following the agreement, Union Minière reopened the Shinkolobwe Mine with equipment furnished by the United States and Britain.[35]

Uranium extraction occurred at great human cost due to early mining techniques, poor health and safety standards, and rudimentary knowledge about radioactive exposure. Congolese workers at the Shinkolobwe Mine received minimal wages and worked around the clock to meet U.S. demands. Through the remainder of the war, several hundred tons of uranium were shipped monthly from Africa to the United States. However, the presence of German U-boats made the Atlantic crossing dangerous. Two of forty ships carrying uranium were lost before reaching their port—one through enemy action and the other through a marine accident.[36]

Secret Cities

Initially, scientists thought that atomic weapons could only be created from uranium-235; however, by the middle of 1941 Glenn Seaborg had determined that plutonium-239 was also fissionable. Unsure of the best choice, the Manhattan Project pursued both simultaneously by constructing three "secret cities," each populated by 125,000 people and absent from maps during World War II. These cities—referred to only by their code names X, Y, and Z—were in remote areas. Today they are known as Oak Ridge, Tennessee; Hanford, Washington; and Los Alamos, New Mexico.[37]

After uranium oxide ore arrived in the United States, it was shipped to St. Louis, where Mallinckrodt Chemical Works operated the country's only production plant for uranium from 1942 to 1951.[38] Once purified, the uranium was sent to Oak Ridge, Tennessee, for enrichment as uranium-235. Oak Ridge also had a pilot plutonium plant that informed the work at Hanford, Washington, where scientists and workers were engaged in plutonium production. Both the uranium and plutonium went to Los Alamos, New Mexico, where a bomb research development laboratory constructed the secret weapons.[39]

Beyond these locations, secret activities related to the Manhattan Project occurred in many existing U.S. cities, including New York; Washington, D.C.; Wilmington, Delaware; Chicago; Boston; Rochester, New York; Berkeley, California; Pittsburgh; Cleveland; Detroit; Dayton, Ohio; Montreal; and London.[40]

The Workers

At the Mallinckrodt plant, employees knew they were working with uranium but didn't know how it would be used. Upon hearing uranium was radioactive, one early worker concluded it was probably used for radios. Although personnel initially used the term "uranium" in describing their job, such transparency drew the concern of military police. Company executives created the name "Uranium Oxide S. L. 42–17" to give the impression that the uranium was just another Mallinckrodt chemical. After attending company lectures on the importance of secrecy, workers used the term "tube alloy" in place of "uranium." The code names of "biscuit," "juice," "oats," "cocoa," and "vitamins" described various stages of the purification process—causing some observers to say the company's correspondence looked like a breakfast menu.[41]

One employee missed the company lecture on secrecy and mentioned in a nearby bar that he was working on uranium at the plant. Five hours later, FBI agents were "all over the bar." They found the man within a day and made sure he didn't talk about uranium again. Another worker learned FBI agents had been quizzing his neighbors about his character. Did he gamble? Chase women? Drink heavily? Was he rowdy? Did he talk about his job? Still another uranium worker was approached by FBI agents. "What did I do?" he asked. "Nothing," they responded. The agents just wanted the man to keep his eyes and ears open and contact them if he heard anything suspicious.[42]

In 1942 there were 24 employees in Mallinckrodt's uranium division, a number that would grow to 1,050 in the early 1960s. (A photograph of 13 Mallinckrodt uranium workers at a mid-1950s safety awards ceremony shows they were all White men.) The early uranium workforce received favorable wages—75 cents per hour as compared to the prevailing 60 to 70 cents. They maintained a common bond with one another and had a sense that their job was important. Before long they would learn how important.[43]

Enola Gay Meets Little Boy

Far from St. Louis, Lieutenant Colonel Paul W. Tibbets Jr. assumed command of the newly created 509th Composite Group of the Army Air Forces on December 17, 1944. Its top-secret mission was to drop the world's first atomic bomb. That purpose was so closely guarded that Admiral Chester W. Nimitz, commander in chief of the Pacific Theater, was unaware of the atomic bomb until February 1945. After training in Utah, the 509th moved to Tinian Island in the Marianas in May, June, and July.[44] During this period the unit's members received the welcome news that on May 8, 1945, the war in Europe had ended in Allied victory.

In conjunction with the Manhattan Project, the U.S. Army detonated the world's first nuclear device on July 16, 1945. Made of plutonium and nicknamed "Gadget," it exploded at what was then the Alamogordo Bombing and Gunnery Range in New Mexico. Physicist J. Robert Oppenheimer, director of the Los Alamos Laboratory, had given the operation the code name "Trinity." After viewing the enormous fireball from the successful blast, he recalled a quote from the *Bhagavad-Gita*: "Now I am become Death, the destroyer of worlds."[45]

Ten days later, the United States, China, and Great Britain issued the Potsdam Proclamation calling for Japan's immediate, unconditional surrender. Although the signatories warned that the alternative was "complete and utter destruction," Japan rejected the proclamation within three days. On the morning of August 6, Colonel Tibbets and eleven crew members left Tinian Island in the *Enola Gay* airplane. They dropped the first atomic bomb, called "Little Boy," on the Japanese city of Hiroshima. Weighing approximately 9,700 pounds, the explosive contained uranium-235 from the Belgian Congo that had been refined at the Mallinckrodt Chemical Works in St. Louis.[46]

Newly in office after the death in April of Franklin D. Roosevelt, President Harry S. Truman issued a statement to inform the American public of the Hiroshima bombing. (Mallinckrodt workers didn't know until the announcement that they had been engaged in building an atomic bomb. Before becoming president, Truman, too, had been kept in the dark about the secret weapon.) He noted that the explosive dropped on Hiroshima was the largest in the history of warfare, possessing more power than twenty thousand tons of TNT. In addition to crediting scientific breakthroughs, as well as financial and industrial resources that made the bomb possible, Truman acknowledged the contributions of workers who labored in defense and production facilities for two-and-one-half years. The president stated, "Few know what they have been producing." They saw "great quantities of material" going into plants and "nothing coming out," since "the physical size of the explosive charge [was] exceedingly small." Emphasizing the protection of the health of workers in the Manhattan Project, Truman said, "They have not themselves been in danger . . . beyond that of many other occupations, for the utmost care has been taken of their safety." He summed up the achievement by declaring, "We have spent two billion dollars on the greatest scientific gamble in history—and won."[47]

A flood of radio announcements and newspaper articles communicated the shocking message that a massive new weapon had been used against Japan. Subsequent news reports continued to stun the American public. On August 8 the Soviet Union (a U.S. ally in Europe) declared war on Japan. The following day a second atomic bomb ("Fat Man") was dropped on the Japanese city of Nagasaki. Containing plutonium-239, it had a different composition from Little Boy. On August 14 Japan surrendered, thereby ending World War II.[48]

At the Mallinckrodt plant in St. Louis, uranium workers were elated. They got a day off work—for some, only the second or third day in as many years. These employees also received a certificate from Secretary of War Henry Stimson and a silver medal the size of a nickel bearing an "A." The certificate, given "in appreciation of effective service," indicated they had "participated in work essential to the production of the atomic bomb." The personnel, especially those who had received a 4-F classification or work-related deferment, were proud to have done their part in winning the war. A case in point was Larry

Faulkner, whose asthma disqualified him for World War II military service. Faulkner later recalled, "I felt like I was doing something. My brother was taken prisoner in Germany. Two of my brothers and my nephew were decorated. My son served in Vietnam. All I can say is, 'I worked for Mallinckrodt.'" As the country celebrated the war's end, there was no expectation that the health of the uranium personnel would be a matter of future concern. A plaque commemorated the achievement on Building 51 of the Mallinckrodt plant, reading, "In this building was refined all the uranium used in the world's first self-sustaining nuclear reaction December 2, 1942."[49]

In addition to Mallinckrodt, St. Louisans recognized other local industries that contributed substantially to U.S. victory in World War II. Aircraft companies at Lambert–St. Louis Municipal Airport produced 3,000 military planes.[50] In the northwest portion of the city, workers at the St. Louis Ordnance Plant made 6.7 billion .30- and .50-caliber cartridges for rifles and machine guns. Beginning production only 9 days after the Pearl Harbor attack, at its peak the plant employed 35,000 people in 300 buildings and bunkers. Half of the employees were women, and the workforce was organized in shifts that kept production going 24 hours a day, 7 days a week. When the plant employed African Americans solely for janitorial and support work, 300 Black protesters marched on June 30, 1942, seeking better-paying production jobs. Managers initially agreed to introduce a segregated production line; however, production became fully integrated in December 1944 at the federal government's insistence (and over the objections of some White workers).[51]

Second Thoughts

Eventually some U.S. citizens had second thoughts about the way the war ended. American George Weller was the first foreign journalist to enter Nagasaki after the atomic bomb had dropped there. He reported how Japanese doctors were puzzled by an undiagnosed "atomic illness" that was killing patients who outwardly appeared to have escaped the bomb's effects. Weller's stories were censored and considered lost until his son published them in 2006.[52]

Government efforts to obfuscate the effects of radiation began immediately after the dropping of the first atomic bomb, according to recent reports in the *New York Times*. Three months after the bombing of Hiroshima, General Groves told Congress that succumbing to radiation was "a very pleasant way to die." He wanted the atomic bomb to be viewed as a deadly form of traditional warfare rather than a new, inhumane type. Since biological and chemical weapons had been banned by international treaty since 1925, Groves wanted "no depiction of atom bombs as uniquely terrible, no public discussion of what became known as radiological warfare." Unfortunately, *Times* science reporter William L. Laurence advanced the government's propaganda while supplementing his newspaper salary through payments from the Manhattan Project. Laurence's coverage was countered by Black war correspondent Charles H. Loeb, who described how "bursts of deadly radiation had sickened and killed" Hiroshima's residents. Loeb's articles were distributed across the United States by the National Negro Publishers Association.[53]

Gradually Americans received reports on the full effects of the atomic blasts. John Hersey's book *Hiroshima* formed the entire August 1946 issue of the *New Yorker*.[54] While earlier accounts had focused on the physical effects of the bomb, Hersey portrayed its impact on six survivors in "haunting and personal" detail. Initially, 85 percent of Americans had approved of the use of the atomic bombs that caused an estimated 140,000 deaths by the end of 1945.[55] However, one year later citizen confidence in the decision was waning.[56] In the face of public criticism—and at the urging of Harvard president James B. Conant—Secretary of War Henry Stimson offered the American people his rationale for using the atomic bomb. He cited the Japanese refusal to surrender, as well as estimates that an Allied invasion of Japan would have cost one million American casualties and many more Japanese ones.[57] In the years that followed, Truman's decision to use the atomic bomb continued to spark interest and controversy among journalists and scholars alike.

In 1999 the Newseum, a museum of the news media in Arlington, Virginia, asked sixty-seven American journalists to rank the top one hundred stories of the twentieth century. They put the atomic bombing of Japan in first place—ahead of the moon landing, the attack on Pearl Harbor, and the first successful flight of the Wright brothers. The subject also attracted scholars who published a wealth of literature on Truman's decision, creating considerable "polarization and acrimony," according to historian J. Samuel Walker. The controversy "reached its zenith—and its nadir—" as the Smithsonian Institution planned a fiftieth anniversary exhibit on the *Enola Gay* and the end of World War II.

The fundamental issue dividing scholars was whether use of the atomic bomb was necessary to win the war in the Pacific on terms acceptable to the United States.[58]

Traditional scholars thought Truman faced a binary choice between dropping the bomb and invading Japan due to the country's refusal to surrender. They also contended that an invasion would have cost the lives of huge numbers of Americans.[59] However, revisionist academics (who rose to prominence in the mid-1960s) argued the bomb was not necessary to win the war because Japan was "teetering on the edge of defeat" and close to surrendering in August 1945. According to this view, the Japanese were seeking a way to exit the war on the sole condition that their emperor be allowed to remain on the throne. Revisionists also wrote that, while Truman and his advisers were aware of the situation in Japan, they still decided to drop the bomb for diplomatic rather than military purposes—primarily, to impress the Soviet Union in the emerging Cold War.[60]

J. Samuel Walker reports that the revisionist and traditionalist camps "were more adept at exposing flaws in the arguments of their adversaries than in providing a convincing answer to the crucial question of whether the use of the bomb was necessary to achieve a timely victory over Japan."[61] Accordingly, during the 1990s, scholars of "a middle-ground persuasion" offered fresh perspective on the subject and a new synthesis. They generally agreed with the traditionalists that Truman used nuclear weapons to shorten the war and save American lives; however, they rejected the notion that the president faced a clear choice between the bomb and an invasion. "With varying degrees of certitude," they held that the war would likely have ended before the invasion of Japan became necessary. Several expressed doubts that American casualties would have been nearly as large as Truman and other officials claimed.[62]

Scholars in all three camps who estimated the casualties from Little Boy generally ignored the prospect of *civilian* deaths in the United States. However, seventy years later, some people in the Coldwater Creek watershed linked the atomic bomb to the illnesses and deaths in their region. For example, registered nurse Mary Oscko regularly exercised and never smoked yet received a diagnosis of stage-four lung cancer. Although unaware of the Manhattan Project until 2013, Oscko believed there was a relationship between the wartime processing of uranium in the St. Louis area and the number of serious illnesses in her neighborhood. Well before Oscko's illness, other "casualties" emerged among Mallinckrodt employees who worked with uranium during and after the war.[63]

Enter the Cold War

The Manhattan Engineer District officially ended in 1946 and became part of the newly created Atomic Energy Commission (AEC). In 1945, Arthur Holly Compton became chancellor of Washington University in St. Louis, a position he held until 1953. The Mallinckrodt Chemical Works continued to refine large quantities of uranium to meet the U.S. government's demands for weapons in the Cold War. It also experimented with thorium.[64]

In 1947, Mallinckrodt hired engineer and former marine Mont Mason to assist in protecting employees from radiation and hazardous chemicals. Some employees dozed off during his lectures; others laughed when a foreman suggested the work might make them sterile.[65] "We were young, just back from the war, and Mason and these guys were talking about protons and neutrons," one recalled. He added, "A lot of us didn't understand what they were talking about."[66]

Regarded as a straight shooter, Mason told workers that while scientists didn't think radiation would be a problem, no one could be certain of what it would do to them. Until his death in 1988, Mason pleaded with government officials for studies on uranium workers' health. He was infuriated that the United States never completed definitive studies on how the allowable radiation levels in the nuclear industry affected people. In fact, uranium is naturally radioactive. Its nucleus is unstable, so the element is in a constant state of decay, seeking a more stable arrangement. It becomes more dangerous as disintegration occurs.[67]

Between 1942 and 1947, most of Mallinckrodt's uranium processing occurred with no limits placed on the amount of radiation exposure for employees. (The *St. Louis Post-Dispatch* reported that early Mallinckrodt workers experienced daily levels of radiation dust over two hundred times the acceptable limits in 1989, the year the story was published.) However, the company did take some precautions. Beginning in 1942, Edward Mallinckrodt Jr. insisted that employees wear respirators and undergo tests at Barnes Hospital in St. Louis. It was a step most nuclear plants didn't take until four years later. Moreover, workers who handled pitchblende were required to shower before lunch, before going home, and anytime they became dusty. In 1945 Mallinck-rodt employees started wearing badges to measure radiation. Then in 1950, after working with the government to establish exposure limits, the company transferred thirty-six people with the highest cumulative exposures out of the uranium division. They were told there was no cause for alarm; however, their exposure was high enough to make it unwise to continue that work.[68]

CHAPTER 2

The Deposit

My God, I didn't know that site was so close to the airport.

—George H. W. Bush, president of the United States, 1989–93

On September 6, 1946, the *St. Louis Post-Dispatch* ran a story with the headline "Refuse from Atomic Ore Stored Here— Condemnation Suit Reveals War Use of Tract near Airport." One might expect readers to be stunned by that disclosure; however, the article drew little notice due to its assurances of safety. The *Post* explained that officials "refused 'for security reasons' to disclose the exact nature of the stored materials, but they declared they [were] not radio-active and not dangerous." The article further described the refuse as "the type . . . that any ordinary commercial firm . . . would store there." Since there was a "remote possibility of [its] future use," the ore was "being stored," necessitating government ownership of the land.[1]

This information was false on several counts. The federal government was storing byproducts of Mallinckrodt's uranium refining for the Hiroshima bomb. The material was radioactive, dangerous, and unsuitable for any ordinary business. Keeping it at the airport enabled a Belgian company to process the material for metals of commercial value. (Fifty-four years later, the *Post-Dispatch* reported that officials had misinformed the newspaper about the storage site.)[2]

As early as March 2, 1946, the Manhattan Engineer District obtained consent to use a 21.7-acre site in north St. Louis County for Mallinckrodt's uranium residues and process waste. Then on January 3, 1947, the property—subsequently called the St. Louis Airport

Storage Site (SLAPS)—was acquired by condemnation in a suit filed in federal court on behalf of the War Department. The cost of the acquisition was $20,000.[3] Located immediately north of the municipal airport, the land was bounded on the south by the Wabash Railroad tracks, on the north by Brown Road (now McDonnell Boulevard), and on the west by Coldwater Creek.[4]

What Change?

Well before the lawsuit's conclusion, trucks had begun delivering radioactive residues and wastes to the SLAPS site. However, the property and byproducts were no longer under the control of the Manhattan Engineer District. The Atomic Energy Act, passed by Congress in 1946, replaced the MED with the Atomic Energy Commission. The new entity had broad powers to conduct, control, and regulate atomic (nuclear) research for military and civilian purposes. By January 1, 1947, the MED was required to transfer to the AEC all materials, facilities, equipment, items, and property related to atomic research.[5] As a result, the AEC supervised SLAPS from 1947 to 1953, and then contracted with Mallinckrodt to operate it until 1967.[6]

In a study of atomic governance in the post–World War II period, Mary D. Wammack finds, "The Manhattan Project was not replaced—it was absorbed." With former Manhattan Project administrators "husbanding the AEC program from the inside out," and with Manhattan Project–affiliated officers embedded in key positions, "a focus on weapons development and the monopolization of production facilities and resources was all but guaranteed."[7]

Many aspects of MED culture remained in the new agency. They included top-secret operations, false safety assurances to the public, the prioritization of nuclear production over the environment or public health, and a practice of hiring contractors while not closely overseeing them. Importantly, one AEC shared value was a lack of interest in nuclear waste. Carroll L. Wilson, the commission's first general manager, said the subject had little appeal among chemists and chemical engineers because it "was not glamorous; there were no careers; it was messy." Wilson added, "Nobody got brownie points for caring about" waste. As a result, the AEC neglected nuclear waste over the years because "there was no real interest or profit in dealing with the back end of the fuel cycle."[8]

Wammack's findings and Wilson's comments support the work of theorist Edgar H. Schein. He describes *organizational culture* as a pattern of shared assumptions that have been accepted by a group of individuals as they solve problems. Because the assumptions have worked in the past, group members convey them to new people as standard ways of thinking about, perceiving, and approaching future problems. In this manner, the values of the Manhattan Project's crash wartime effort continued under the Atomic Energy Commission even though the war was over.[9]

In 1947 the AEC appointed a committee of environmental, safety, and health experts to assess the effects of the radioactive and toxic materials used in weapons production. Panel members reported that the disposal of contaminated waste, "if continued for decades," presented "the gravest of problems." They added, "This is one of the areas of research that cannot be indefinitely postponed." Nevertheless, the AEC ignored the recommendations, and four decades passed before the federal government began to seriously address the consequences of placing weapons production ahead of environmental concerns.[10]

During the entire time the AEC and Mallinckrodt operated SLAPS, it received residues and wastes from the company's downtown plant. These substances included pitchblende, raffinate, radium-bearing wastes, barium sulfate cake residue, and Colorado raffinate residues, among others. Some of the byproducts were stored in fifty-five thousand thirty- and fifty-gallon drums; they were packed by Mallinckrodt employees and extended end to end over most of the site.[11] Other byproducts were stacked on the open ground. One report notes that, for barium sulfate alone, the cake pile was "25 feet high and covered three acres" until 1960. In 1954, SLAPS also received sixty tons of captured Japanese sand containing uranium residues and waste.[12]

Special Delivery

For over a decade, navy veteran Tom Green and four other independent drivers hauled unnamed material along a well-traveled haul route on a twenty-two-mile round trip from the Mallinckrodt Chemical Works to SLAPS. Green alone transported at least five thousand tons of material each year, with each load weighing between eight and nine tons. None of the drivers knew what they were carrying, and the nature of the destination was a secret.[13]

The trip began at the Mallinckrodt plant on Destrehan Street in

St. Louis. Proceeding north on Broadway Avenue through an industrial area, the motorists passed two large city cemeteries. Angling left (northwest) on Halls Ferry Road, they entered neighborhoods of tidy brick homes in north St. Louis and its close-in suburbs in north St. Louis County. Upon reaching Highway 66–67 (now Dunn Road, just north of Interstate 270), the truckers drove west across agricultural land that would house two large high schools. They turned east on Brown Road (now McDonnell Boulevard) to reach the St. Louis Airport Storage Site (SLAPS) by the municipal airport.[14]

Spillage from delivery trucks was a common occurrence, and the public remained unaware of the risks involved. For example, on March 2, 1953, the *St. Louis Post-Dispatch* carried an article titled "Radioactive Dirt on Highway Came from Chemical Works." The story describes a dump truck accident at what was then the Highway 66 bypass, two miles north of the airport. The driver stated that when the car ahead of him suddenly slowed down, he swerved and overturned in a ditch, spilling his cargo. While reporting that the upset material came from a project of the Atomic Energy Commission, the *Post-Dispatch* communicated official assurances that "the dirt was not dangerous." Following the accident, the dirt was washed from the road by the all-Black volunteer fire department of Robertson, Missouri, "only because it was a hazard to motorists."[15]

Upon arrival at the landfill, Green's cargo often adhered to his shoes, and winter snow turned the mixture into a quagmire. His vehicle would slip and slide, sometimes becoming stuck in "the muck." Green's health and exposure were never monitored. Although he would eventually learn that the material was radioactive, Green was not concerned during his twelve years of hauling what he called "the richest dirt in the world." However, after getting cancer, he said his job might have cost him his life. Tom Green died on June 8, 1979, at the age of sixty-three. His death certificate attributes the cause to cancer of both lungs. He had smoked cigarettes but stopped several years before his death.[16]

At the airport site, there were no apparent controls for groundwater, surface water, or exposure pathways. The drivers deposited many deliveries in piles as the acreage—previously a patchwork of farm fields—became a "moonlike world." Mallinckrodt workers circled the area in bulldozers and trucks, reshaping the earth to make room for more waste. One employee, Richard F. Schroeder, later recalled his enjoyment at making mountains, moving them, and carving out mesas

St. Louis Airport Storage Site, circa 1958.
U.S. ARMY CORPS OF ENGINEERS

and roads. Occasionally the workers drove their cars to the top of piles some forty feet tall to watch planes fly in and out of the airport. Bruno Bevolo remembered one instance when, at the direction of the Atomic Energy Commission, he and his colleagues buried a pickup truck that had become "too hot" for further use.[17]

A chain-link fence surrounded the area, where guards were present from 1946 to 1951. In time, SLAPS had a small building containing showers, a change room, and office space. In 1959 the site added railroad-loading facilities and a "siding" (a low-speed track section distinct from the through route).[18]

The Creek

Coldwater Creek (known to French settlers as Rivière de L'eau Froide) is a minor tributary in one of the world's great river systems. It includes the Mississippi River and its major tributaries—the Illinois, Missouri, Ohio, and Arkansas). Together they drain one-third of the U.S. landmass from the Appalachians to the Rocky Mountains. Coldwater Creek begins at the present site of Overland, Missouri, and flows (by gravitational pull) north and east for nineteen miles. It empties into the Missouri River three miles west of its confluence with the Mississippi.[19]

Choosing the SLAPS site for the storage of radioactive material was problematic. Its terrain was uneven, with a low drainage area in its western section. One-third of the site was in a floodplain. Moreover, the land had an east-to-west slope with all surface water and ground-water draining to Coldwater Creek.[20] When wind, rain, snow, and gravitational pull eroded the piles of radioactive waste, they drained into the creek. As some barrels began to rust and decay, their contents spilled to the ground and joined the residue already there.[21] As radio-active waste leached into Coldwater Creek, it continued its remaining 13.7-mile journey through north St. Louis County before emptying into the Missouri River.[22] Some radiological material entered the groundwater through wells near the deposit site, while other pollutants continued downstream. Just northwest of the creek's mouth was a sinkhole (karst) area overlaid by loessial soils. According to the Army Corps of Engineers, the area exhibited "internal draining directly into the groundwater system."[23]

To make matters worse, the area was subject to flooding. By definition, a small stream, or "creek," seasonally floods and is rich with sediment.[24] A local history observes that Coldwater Creek was "ordinarily the best behaved of streams." However, "at flood-stage," it carried a volume of water that often took its toll "in serious damage to adjoining fields and houses."[25] During these times radioactive contaminants would spread to wider areas beyond the creek's banks. Although in 1946 SLAPS was in a predominantly rural locale, the population would grow in the postwar period, and the airport would become busier, potentially spreading contamination to more people.

Questions and Answers

Aside from the environmental problems inherent in the SLAPS selection, there were other issues that, in time, aroused citizens' concern. Some, including the forty-first president of the United States, were aghast to learn the location of massive amounts of radioactive waste. People asked, "Why did they dump it by the airport?" A short answer would attribute the action to politics and a limited understanding of land pollution. However, a meaningful response requires answers to several additional questions: Why didn't officials dispose of the radio-active material at the Mallinckrodt plant? How did the airport enter into the decision? Why was there no concern for Coldwater Creek's downstream population?

Before 1946, the residues and waste from uranium purification remained at the Mallinckrodt Chemical Works, a forty-five-acre manufacturing facility located three hundred feet west of the Mississippi

River.[26] In 2019, Susan Adams, project engineer for site cleanup efforts, acknowledged that early uranium processors simply dropped waste down the drain. She added this action had occurred before the passage of key environmental legislation and, unfortunately, was "normal practice" at the time. As a result, the soil around the Mallinckrodt plant became contaminated through leaky sewer systems and flooding.[27]

The Mallinckrodt factory was further affected by the U.S. promise to purchase only uranium in the pitchblende and let African Metals Corporation retain the radium and other precious metals contained in the ore. While the agreement gave the Manhattan Project immediate access to large quantities of uranium, Mallinckrodt Chemical Works soon ran out of storage space at its St. Louis plant.[28]

In seeking a solution, officials chose a property in a sparsely populated area adjacent to the airport. Although the site's railroad access might have favored its chance of selection, the topography was less than ideal. Moreover, the condemnation suit was opposed by co-defendants Elizabeth Callaway and Mary Callaway Porcher (landowners), Draining District No. 2A, St. Louis County, and the St. Louis County collector of revenue.[29]

The SLAPS site had an interesting past. The owners inherited it from prominent family member James C. Edwards, president of the North Missouri Railroad (a predecessor of the Wabash). During the Civil War the Union army jailed Edwards in St. Louis on suspicion of being a Confederate sympathizer. In 1863 he signed a loyalty oath to the United States and later became an administrative judge in St. Louis County following its separation from the City of St. Louis.[30] However, it was not Edwards's loyalty to the United States, but rather the separation of St. Louis City and County that encouraged the placement of Mallinckrodt wastes on the SLAPS property. The schism was one of several factors influencing the site decision.

City versus County

St. Louis's secession from St. Louis County occurred in 1877 following a referendum the previous year. St. Louis voters favored forming two separate political entities and extending the city's municipal footprint by forty square miles. With that acquisition, St. Louis's landmass was frozen at sixty-one square miles and blocked from further expansion.[31]

The secession was largely prompted by the differences between the area's urban and rural residents. Over three hundred thousand people lived within the city, while only thirty-one thousand populated the rest of St. Louis County. City dwellers characterized the St. Louis County Court as remote, inefficient, and corrupt. They expressed dissatisfaction after 1867 when the Missouri legislature gave the county court the power to assess and collect taxes.[32] Moreover, residents of the city did little to disguise their condescension toward their rural neighbors, balking at paying for the roads, water supplies, sewers, and other facilities the county would require.[33] In what became known as the "Great Divorce," the Missouri legislature separated the townships of St. Ferdinand, Central, Bonhomme, Meramec, and Carondelet from the City of St. Louis and formed a new county of the same name, with Clayton as its seat.[34]

At the time of the Great Divorce, St. Louis was one of the largest manufacturing cities in the United States, and it "generated more than its share of industrial pollution," according to historian Andrew Hurley.[35] In studying the regulation of nuisance trades from 1865 to 1918, Hurley finds that St. Louis routinely quarantined toxic industries, "usually in areas where political opposition was weakest, property values were lowest, and residents were poorest."[36]

Hurley points to the actions of the St. Louis board of health, established in 1867 with broad powers to abate public nuisances in the industrial city. One city council member suggested buying land outside the municipal boundaries for the sole purpose of housing nuisance industries. However, the idea failed to gain traction because health officers would lack jurisdiction to hold errant manufacturers accountable. (Hurley explained, "On occasions when winds blew fumes back toward the city, the consequences might be unbearable.") The board instead asked the city council to establish special areas within St. Louis where nuisance industries could congregate. When the council refused, the board began the *selective* enforcement of nuisance laws. Members hoped polluters would avoid areas where laws were strictly enforced and congregate in others where enforcement was lax.[37]

Hurley notes that St. Louis "followed the physical contours of the landscape" by situating its wealthiest neighborhoods on high elevations with good air circulation, pleasing vistas, and ground "best protected from unforeseen floods."[38] The board of health therefore opposed locating a nuisance district in the city's western neighborhoods, due to the negative effect on property values that were among the highest in St. Louis.[39] Instead, the board insisted that nuisance industries find "a more suitable location," one "settled almost entirely by the poorer classes." Seeing the irony, Hurley states, "Begun under the auspices of protecting the general public from disease, pollution control increasingly became a tool for protecting private property and thus, distinguishing the environmental experiences of rich and poor, powerful and powerless."[40] In 1918, St. Louis became the second major U.S. city, after New York, to devise a comprehensive zoning law with land-use restrictions. Special districts known as unrestricted areas were set aside for manufacturers producing objectionable noise and emissions.[41]

The Airport's Appeal

Both St. Louis's Great Divorce and its quarantining of noxious industries contributed to a decision to "store" radioactive wastes by the airport. Owned by the city and surrounded by a different political jurisdiction (St. Louis County), the airport was advantageously located far from city voters. In 1946, there was one aircraft manufacturer at the airport, and more industry could be expected there in the future.

Although the City of St. Louis was prohibited from further expansion, its airport was not—and it continually enlarged its footprint. The "St. Louis Flying Field," as it was first called, was the creation of Albert Bond Lambert, a licensed airplane pilot who led a family pharmaceutical company known for selling Listerine mouthwash. In 1920 Lambert and the Missouri Aeronautical Society obtained from Mary Jane Weldon a five-year lease on 170 acres of farmland in Bridgeton. Its flat lake-clay surface was once an immense prehistoric lake. Lambert made the payments and arranged for the site to be cleared, graded, and drained. He also constructed a hangar at his own expense and invited any aviator to use the facility without charge.[42]

When the lease terminated in 1925, Lambert bought the property and two years later offered it to the City of St. Louis for use as one of the first municipal airports in the United States. The proposal was contingent on the passage of a bond issue that city voters approved five to one, seemingly energized by Charles Lindbergh's 1927 flight from New York to Paris. (The plane's name, the *Spirit of St. Louis*, recalled the backing of St. Louis businessmen, including Lambert.) As he advocated for the bond issue, Lambert leased the proposed airport land to the city for one dollar and purchased an adjoining tract of seventy-six acres for future expansion and development.[43] On July 12, 1930, area residents attended the dedication of the Lambert–St. Louis Municipal Airport. Beyond being recognized through the facility's name, Lambert received an appointment to the St. Louis Airport Commission, a group that engaged consultants and made recommendations on municipal airport sites and improvements for many years.[44]

As the airport undertook expansions, Lambert warned radio audiences that St. Louis risked falling behind Chicago, Detroit, Cleveland, and Kansas City if it did not increase air travel. City residents still smarted from St. Louis's loss to Chicago as the primary railroad hub in the nation's center. However, Lambert argued that St. Louis had a future as "the aerial crossroads of America."[45]

When World War II ended, the Lambert–St. Louis Municipal Airport comprised 1,060 acres, a six-thousand-foot runway, U.S. Navy and National Guard bases, and an aircraft factory.[46] Yet despite (or perhaps because of) the growing interest in air transportation, the airport struggled. By 1946, airline flights and passenger volume had reached three times the 1941 levels. Commercial flights vied with private aircraft concerns and navy traffic in the air and on the ground. The terminal building (constructed in 1933) was inadequate for the demands of the postwar era.[47] When the federal government acquired the SLAPS site, it was a safe bet that the airport would continue to extend its footprint. It was also likely that the United States would give SLAPS to the City of St. Louis after the "stored" material was returned to the Belgians.

An Inconvenient Waterway

Throughout its expansion, the Lambert–St. Louis Municipal Airport faced two significant problems: flooding from Coldwater Creek on its eastern boundary, and jurisdictional issues with St. Louis County. In resolving these concerns, the city sometimes solicited the help of the federal government.

To address the problem of flooding, St. Louis officials embarked on a series of steps to control the pesky creek that affected the airport. The U.S. government proved to be a useful partner in that endeavor. In 1930, the city authorized $275,000 to reroute Coldwater Creek to facilitate runway construction.[48] However, when officials decided to extend the northeast–southwest airport runway from 3,200 to 4,500

feet, the creek was again in the way. Agreeing that a further relocation of the tributary would be impractical, St. Louis officials persuaded
the Works Progress Administration (WPA) of the federal government
to fund a culvert to carry the creek under the runway extension. This
was the first of two instances of the airport's bridging over Coldwater
Creek.[49]

Federal officials were also helpful in resolving jurisdictional issues
between St. Louis City and County. In the late 1930s, aviation pioneer
Jimmy Doolittle chaired a committee to identify needed improvements
at Lambert–St. Louis Municipal Airport. The group recommended en-
closing Coldwater Creek and removing (along with other obstructions)
a sixty-five-foot grain elevator that stood in direct line with the end of a
runway. Despite the danger it posed to air travel, the grain elevator remained in place until 1944. Its owners refused to sell, but the city could
not exercise eminent domain to acquire the structure because it was in
St. Louis County. During World War II the city finally persuaded the
federal government to use its power of eminent domain to acquire the
grain elevator and remove it.[50]

In 1944 the St. Louis Airport Commission recommended purchas-
ing 350 acres on the southeast side of the airport property as the site of
a terminal and runways. However, expanding to the southeast entailed
dealing once again with Coldwater Creek. It would be more than ten
years before Lambert airport opened a new passenger terminal.[51] By
the war's end, the portion of the creek that flowed on airport property
had largely been covered—a feat that shaped perceptions of the tributary. Officials viewed it as being in the way or in need of control. Once
covered, the Coldwater Creek was out of sight and out of mind.

Views of Pollution

Twenty-first-century readers may find it implausible that Mallinckrodt uranium processors had no qualms about putting radioactive
waste down the drain near the Mississippi River. Readers may be more
surprised to know that—while officials were piling radioactive materials on the SLAPS property—St. Louis was a national model for air
pollution abatement. Exploring this apparent disconnect illuminates
how St. Louisans viewed pollution and its relationship to the air, water, and land.

After long being regarded as the "dirtiest place in the Mississippi
valley," St. Louis had a series of failed attempts to abate an ongoing

smoke problem caused by a reliance on bituminous coal for its primary fuel.[52] The situation was particularly severe in the autumn of
1939, when residents couldn't see across their thoroughfares. Under
the leadership of future mayor Raymond Tucker, the city organized
a comprehensive plan for smoke abatement. Championing smokeless
fuel and proper burning equipment, a citizens' committee consisting
of over one hundred civic groups, elites, and the press brought pressure for the successful passage of a smoke control law. Tucker later recalled that it changed "the buying habits of approximately 1,000,000
people . . . [and] the merchandising habits of practically all the fuel
dealers in the city of St. Louis." The results were dramatic. The weather
bureau estimated that the hours of "moderate" smoke decreased by
70.3 percent, and those of "thick" smoke decreased by 83.6 percent.
By 1944, 230 municipalities had requested information about St. Louis's smoke ordinance.[53] In time, coal lost out completely to natural gas
and oil as domestic fuel, and the diesel-electric locomotive replaced
the smoky steam locomotive.[54]

Given this successful case of environmental advocacy, one wonders
how the same city could have tolerated the presence of radioactive material at SLAPS, bordered by Coldwater Creek. One reason is the secrecy that prevented most people from knowing about the presence of
radioactive material (while they clearly knew that smoke was blocking
their view across the street). Information that did leak out was countered by incorrect official denials, as in the September 6, 1946, and
March 2, 1953, *Post-Dispatch* articles.[55]

In *The Search for the Ultimate Sink*, Joel Tarr explains that U.S. professionals "began to pay greater attention" to the health and environmental damages of industrial wastes only after World War II. Not until
the 1970s did these wastes receive "fuller policy recognition"—and even
then, the major focus was on water and air pollution. (He adds that an
increased concern with land pollution developed in 1965 and grew following the 1970s.)[56] Tarr's assertion is supported by the passage dates
of important environmental legislation. They include the Federal Water Pollution Control Act (1948), Clean Air Act (1963), Motor Vehicle Air Pollution Control Act (1965), Water Quality Act (1965), Clean
Water Restoration Act (1966), Air Quality Act (1967), Clean Air Act
(1970), Water Quality Improvement Act (1970), Solid Waste Disposal
Act (1970), Federal Water Pollution Control Act (1972), Safe Drinking
Water Act (1974), Resource Conservation Recovery Act (1976), Surface Mining Control and Reclamation Act (1977), Soil and Water Re-

source Conservation Act (1977), Hazardous and Solid Waste Management Act (1984), and Superfund Act Reauthorization Amendments (1986).[57]

Tarr's insight helps explain the lack of concern for the land pollution that seeped into Coldwater Creek, harming residents of the watershed. Also noteworthy is the fact that the sparsely populated rural residents of north St. Louis County belonged to a different political entity than the decision-makers. To recall Hurley, these rural people lived "where political opposition was weakest, property values were lowest, and residents were poorest."[58]

Return to Sender?

Since a driving force in SLAPS's creation was the need to "store" the uranium processing residues for African Metals Corporation, readers may wonder whether the massive amounts of radioactive material were ever returned to the Belgians. The historical record suggests that little residue and waste material was sent to African Metals or its parent company, Union Minière du Haut Katanga.

Robert Alvarez, a senior scholar at the Institute for Policy Studies and former senior policy adviser in the Department of Energy, reports that in the late 1940s and early 1950s some 1,157 tons of radium-bearing residues were shipped to a uranium processing plant in Fernald, Ohio, and to the Ontario Ordnance Works in upstate New York. The residues sat unattended in aboveground silos for decades, leaking large amounts of gas into the atmosphere. At Fernald, the red-and-white checkerboard silos (reminiscent of the Purina logo) and name ("Feed Materials Production Center") led many locals to believe the site was making animal feed. The federal government eventually assumed ownership of the material and disposed of it at a cost of $460 million.[59]

Between 1966 and 2000, radioactive residues and wastes were shipped from SLAPS to several destinations within the United States, as discussed in chapter 4. None of the sites belonged to the African Metals Corporation or its parent company. Edgar Sengier remained on the Union Minière board until 1960—the same year Congo became independent of Belgium, and Katanga separated from Congo. Sengier retired to Cannes, France, and died in 1963. Three years later, the Congolese government took over the possessions and activities of Union Minière du Haut Katanga.[60]

Even before Congo became independent of Belgium, there was little apparent interest in SLAPS's "stored" residues and waste. The Kay
Drey Mallinckrodt Collection contains an April 11, 1959, document
from the Atomic Energy Commission (St. Louis Airport Site) that supports this interpretation. It notes the following:

> The Vitro Corporation of Canonsburg, Pa., contracted with the Afri-
> can Metals Corporation to purchase Pitchblende raffinate . . . for re-
> covery of nickel, cobalt, and copper. The [Atomic Energy] Commis-
> sion entered into a contract with the Vitro Corporation to purchase
> uranium values recovered from the raffinate. Since the market value
> of nickel, cobalt, and copper have decreased considerably in the last
> few years, it is understood that Vitro Corporation has cancelled the
> contract. It is our understanding that African Metals Corporation
> may abandon the material in the near future. No plans have been
> made to recover the uranium values of this material.
>
> There are no current plans to dispose of any of the remaining
> materials.[61]

The "History of Material Storage at the St. Louis Airport Storage
Site" also reports, "African Metals abandoned the material following
decreases in the market values of the nickel, cobalt, and copper remaining in the raffinate. The . . . [pitchblende raffinate] was part of
the residue from the site sold in 1966 to Continental Mining and Milling Company, and was moved to the Latty, Missouri, Avenue site from
1966 to 1967."[62]

Despite the disinterest in SLAPS's residues and wastes, there was
support for selling the fifty-five thousand thirty- and fifty-gallon
drums that contained the site's contaminated material. The April 11,
1959, document reports, "[Ten thousand] 30-gallon drums are being
sold." It adds, "The remainder are unsaleable and will probably have to
be baled and sold as scrap metal, together with the 3,500 tons of other
contaminated steel and alloy scrap also stored at the site. It is expected
that procedures will be established at an early date for disposition of
this contaminated scrap metal."[63]

PART TWO

The Dissemination

CHAPTER 3

The Contamination Spreads

I feel like everything was great when you were a child and
then 20 years later you have cancer because of it.

**—Patricia Barry, age thirty-three, former north St. Louis
County resident diagnosed with appendix cancer, 2016**

The radiological contamination of the St. Louis Airport Storage Site
was a cause for concern. Its proximity to Coldwater Creek was even
more alarming. However, these problems were compounded by the
post–World War II housing boom in north St. Louis County. It trans
formed Florissant from a sleepy little town known for exceptional
farmland into the county's largest municipality. As residents remained
unaware of radionuclides in their midst, new construction practices
spread the contamination throughout the Coldwater Creek watershed.

A Bright Future

In 1947, Florissant mayor Arthur Bangert welcomed the opportunity
to promote his city on KXLW radio in St. Louis. He predicted a "bright
future" for Florissant while acknowledging it had been "more or less
dormant" in the past.[1] Historian James Neal Primm concurs with the
mayor's assessment. Primm writes that in 1940 the "old [French] colonial village . . . was still slumbering," having increased its population by "only a few hundred" since the Louisiana Purchase.[2] Yet Bangert had reason for optimism. He pointed to the development of 165
new homes that would increase Florissant's population by 35 percent,
as well as the construction of a $10 million Ford plant nearby. Addi-
tionally, Bangert cited Florissant's advantage of a location "in the St.
Louis shipping zone, . . . nearer the . . . airport than . . . downtown St.
Louis."[3]

Called "the Father of Modern Florissant," Arthur Bangert served as mayor from 1938 to 1950 and is credited with bringing financial stability to the town. Over time, he introduced piped water, a sewer system, a municipal library, and an administrative structure that included the planning and zoning, public works, and police departments. In 1939 he worked with the city council to officially change the town's name from St. Ferdinand to Florissant.[4] Yet despite his efforts to bring the community into the twentieth century, Bangert did not witness its full growth potential. In 1956, he died unexpectedly at age fifty—just as Florissant was on the cusp of a population explosion.[5] Between 1947 and 1980, eighteen thousand homes were built there. Florissant's population would peak at seventy-six thousand in the mid-1970s, making it the largest of St. Louis County's more than eighty municipalities.[6]

Speaking in 1947, Mayor Bangert did not imagine that radioactive material from the first atomic bomb would be stored near the St. Louis airport—or that it would travel downstream to his community. He could not have guessed that standard homebuilding practices would spread radionuclides beyond Coldwater Creek, contaminating Florissant's new residential areas.

Today, however, we know this dispersal of Manhattan Project wastes followed ecological principles. At SLAPS, the force of gravity caused surface runoff, rain, and snow to drain into Coldwater Creek, which unidirectionally flowed from the source to its mouth at the Missouri River. The flow pattern was reversed when the creek backed up into its branches during seasonal flooding, leaving behind rich sediment. Over time, the interaction of these environmental and human factors posed a threat to residents of north St. Louis County.

The World of Tomorrow

The United States experienced a surge in homebuilding in the postwar years that reflected "unprecedented general affluence," according to Richard N. L. Andrews, an environmental policy expert. Since World War II was above all else a war of industrial production, many Americans feared the nation would return to depression and unemployment during peacetime. However, the industries that expanded for war production did not scale back or close in 1945. Instead, they sought new civilian and export markets as well as continued government projects. The heightened production levels of the Cold War led to mass consumption and a general improvement in the U.S. standard of living.

Andrews notes that "not just the wealthy but most households now lived in urban rather than rural areas, and could afford a house, a car, major durable goods and conveniences, recreational trips, and other amenities."[7]

These socioeconomic shifts resulted in dramatic increases in the overall and per capita consumption of environmental commodities, with associated environmental impacts. Between 1900 and 1952 the yearly use of coal increased two-and-one-half times, copper by three times, iron ore by three-and-one-half times, zinc by four times, natural gas by twenty-six times, and crude oil by thirty times. Meanwhile, small automobile engines were replaced by larger, less fuel-efficient ones; air-conditioning came into widespread use, greatly increasing energy consumption; phosphate detergents began replacing soaps; and the use of pesticides heightened dramatically.[8]

As early as 1952 the Commission on Materials Policy (appointed by President Harry Truman and chaired by William Paley of the Columbia Broadcasting System) reported that the United States was depleting its known reserves of strategic materials much more rapidly than other countries. Moreover, the nation's economy had shifted from exporting raw materials to depending on raw materials imports. The commission cautioned, "After successive years of thinking about unemployment, re-employment, full employment, about factory production, inflation and deflation, and hundreds of other matters in the structure of economic life, the United States must now give new and deep consideration to the fundamental upon which all employment, all daily activity, eventually rests: the contents of the earth and its physical environment."[9]

Andrews reports that the commission's recommendations had little effect on the affluent society in postwar America. Until the 1980s, Cold War ideology helped to justify a need to "keep secret" the records of pollution discharges. In some cases, this belief led to "outright deception of the [members of the] public concerning the risks being imposed on them." Andrews adds that the nuclear industry was noteworthy in creating tremendous amounts of toxic and radioactive pollution—and billing the cleanup costs to the next generation.[10]

For many Americans, peacetime offered a welcome respite from the consumer-goods shortages of World War II, when manufacturers could only sell promises. Wartime advertisements imagined "the world of tomorrow" after the conflict ended, according to historian Adam Rome. Seeking to revive faith in a capitalist system shaken by the Great Depression, the ads cited miracles of wartime invention and production that would transform everyday American life in peacetime.[11] Such dreams encouraged a postwar building boom throughout the United States, with an explosion in the suburbs. "World of tomorrow" advertisements had framed the war as a fight for "comfort, freedom, and . . . happiness" while identifying the single-family home at "the centerpiece of a vision for the nation."[12] City dwellers could escape the noise and congestion of urban life and achieve an American ideal: a home on land that could be cultivated, "even if the principal crop were grass."[13]

Earthmovers

The United States was well positioned to encourage postwar homebuilding through troop demobilization, legislation, and infrastructure improvements. The release of 325,000 members of the Naval Construction Battalions ("Seabees")—along with other military construction workers—guaranteed a labor force to meet the pent-up housing demand. Drawing on their wartime experience in large-scale land clearance, some veterans started construction companies or took jobs as heavy-equipment operators. However, it was former Seabee William Levitt who in 1951 built over seventeen thousand houses in the ranch and Cape Cod styles, constructing them en masse on one thousand acres of former potato fields. This community—Levittown, New York—"came to epitomize large-scale postwar residential suburbs."[14]

Federal legislation also aided new home construction. The GI Bill of 1944 built on existing federal mortgage subsidies to give new veterans zero-down-payment loans for new homes. Additionally, the Federal-Aid Highway Act of 1956 encouraged suburban commuting by providing 90 percent subsidies for highway construction. The suburbs received yet another boost in the 1954 Internal Revenue Code, which made it more profitable to build new construction on "greenfield sites" than to reuse existing places.[15] The end result was a significant altering of the U.S. landscape during the postwar years.

In north St. Louis County, environmental factors caused the Coldwater Creek watershed—and Florissant in particular—to be attractive to homebuilders. Situated in "one of the most fertile and valuable prairies in the country," the area's relatively flat land lowered site preparation costs compared with those in south St. Louis County, where woodlands were prevalent and builders had to dig in rock to create basements.[16] Even west-central St. Louis County was a "mosaic of prairie, brush, woods, and timber" as compared with Florissant's landscape.[17]

As construction crews reshaped the landscape, making way for homes, roads, schools, apartments, and shopping centers, residents gave vivid descriptions of earthmovers that "roared through the [Florissant] valley." They were followed by graders that "peeled layers off a ridge near Coldwater Creek," exposing the firepits and projectile points of a Paleo-Indian culture.[18] Most of the area was temperate grassland, also known as "prairie," which—with normal development—has "the world's deepest soils and best agricultural land."[19] However, the heavy equipment turned the rich topsoil under and covered it with clay. The machines also altered Coldwater Creek. Some residents expressed regret at the loss of topsoil; others complained the earthmovers were creating a landscape of sameness, with "one rolling hill after another, with both hills and valleys encrusted with one little house after another." They wistfully recalled the land's former contours: "narrow roads curving through the bald knobs" and "clusters of neat farm buildings . . . surrounded by oceans of black soil, and sometimes by oceans of Missouri River floodwater."[20] Observer Gregory Franzwa described the effect on places that once seemed frozen in time—in this case, a small community in the Coldwater Creek watershed named for the blackjack oaks that stood proudly at the town center:

> Even the little town of Black Jack seems to be on the way out, at least in its rural aspects. The hills, which at one time would have proved formidable to a billygoat, are now being taken in stride by the earthmovers, as apartment houses and new subdivisions continue to move even closer. The ground on those corners which once nurtured those three great oaks, is growing more valuable with each passing day. Maybe it's only a matter of time before the celebrated sign at the junction—"Black Jack—here 'tis—" will be changed to "Black Jack—here 'twas." There are a good many people . . . who hope not.[21]

Place Names

During this period, home construction technology was moving from a craft in which each residence was a distinctive creation to a manufacturing process of "look-alike sets of tens or hundreds of homes." Political scientist E. Terrence Jones notes this practice brought economic efficiencies and lower prices that "broadened the market and generated substantial profits for homebuilders."[22] They chose the names of their suburban developments (subdivisions) to appeal to consumers. However, Rome ruefully observes that in constructing these tracts, builders "routinely destroyed the meadows, woods, and hills they honored in their place names."[23] For example, beginning in 1956, the home-building firm of Mayer-Raisher-Mayer developed Paddock Hills in Florissant, the first among many of its suburban neighborhoods. Since a paddock is an enclosure for keeping or exercising horses, the builders used an equestrian theme to market their homes. Paddock Hills' logo featured a galloping horse, and its winding lanes had names like Saddlespur, Steeplechase, Thoroughbred, and Churchill Downs. After developing this subdivision, the Mayer family continued to create neighborhoods in the Coldwater Creek watershed, often using their successful brand name—as in Paddock Meadows, Paddock Estates, Paddock Woods, and Paddock Forest. As Rome suggests, the geographical features in these names were often missing in real life.[24]

By 1960, Florissant had become "the new middle-income mecca," according to Primm.[25] Advertisements for Paddock Hills emphasized features consumers desired, recalling the "World of Tomorrow" ads of World War II. The amenities included central air-conditioning, built-in gas kitchens, full basements, attached garages, family rooms, covered patios, and aluminum storm windows.[26]

Unfortunately, postwar builders put "hundreds of thousands" of houses in environmentally sensitive areas. They also leveled hills, filled creeks, and cleared vegetation from vast tracts of land. Nature, in turn, responded with more frequent flooding, soil erosion, and dramatic changes in wildlife populations.[27] Rome observes,

> The building boom came at considerable environmental cost. The reckoning of those costs began almost immediately. By 1950, for example, a number of people had already become concerned about the energy-wasting design of the typical postwar home. Yet the critics of tract housing faced formidable obstacles, because the new way of building met so many economic, social, and political demands. The first complaints about the environmental impact of postwar construction seldom made the news. Instead, the media focused on a far more compelling story: The nation's largest builders were answering the prayers of millions.[28]

Environmental Effects

Many scholars have discussed the negative effects of suburbanization. For example, it encouraged commuting, thereby increasing pollution from automobiles; it put homes in environmentally sensitive areas; it caused pressure on the greenbelt and contributed to urban sprawl; it accelerated urban decay and a concentration of lower-income residents in the center city.[29] However, other effects of suburbanization (often largely ignored) put residents of the Coldwater Creek watershed at risk. Earthmovers disturbed radioactive waste that had leached into the soil via the creek, releasing contaminated particles into the air. In the grading that followed, contaminated floodplain soil and creek sediment were moved to serve as "fill" elsewhere. Moreover, the new construction required a significant increase in the amount of land covered by impermeable concrete, dramatically decreasing the amount of water that could be absorbed by the soil and contributing to further flooding. In 1988, the Army Corps of Engineers reported that the creek's one-hundred-year floodplain between Interstate 270 and New Halls Ferry Road—formerly a rural area—was largely covered with single-family homes, apartment complexes, large commercial developments, some industries, and several open spaces. The corps' description appeared as part of a feasibility study for a new flood control project.[30]

Fragmented Government

One result of the population boom in north St. Louis County was the introduction of new local governments. When business and industry moved closer to the airport, Florissant expressed interest in annexing a nearby unincorporated area. In 1944 a group of farmers organized to oppose annexation by a municipality poised for growth—and five years later incorporated as a village.[31] They chose the name "Hazelwood" after a local historic plantation where the daughter of a Confederate general still lived.[32]

The actions of Florissant and Hazelwood exemplify two of three options for structuring local government as St. Louis County grew from a population of one hundred thousand in 1920 to nearly one million by 1970. One option was for all fifteen municipalities existing in 1920 to expand their boundaries to accommodate the growing numbers. While this could be accomplished "with a stroke of the pen," residents in the annexed areas would have no voice in the matter.[33] A second option was for the new suburbanites to choose no municipalities at all. This would require little change since, in 1920, almost two-thirds of St. Louis County residents lived in unincorporated areas. Here, the county could serve as a default supplier of local government services. If it decided not to provide a particular benefit, residents would do without it, purchase it privately, or form a special government district.[34]

The third, and most popular, option was to create new municipalities commensurate with the population growth. During the 1930s, some existing cities began to see the benefits of annexation—especially if they could acquire a wealthier area. Residents who objected could gather signatures from 50 percent of a proposed new city, present the petition to county government, and create a municipality of their own. Between 1945 and 1952 over fifty cities formed in St. Louis County in this manner. In so doing, they sought to protect themselves against outside influences and guarantee their homogeneity.[35] However, such an increasingly complicated patchwork of autonomous government entities could be at a disadvantage in addressing area-wide problems that required cooperation and coordination.

Meanwhile, rural school districts were feeling threatened by the county board of education's proposal to combine all north St. Louis County school districts into one large system. In 1949 the Elm Grove District 9 organized a town district called the School District of Hazelwood. Directors of the new entity worked for two years to amicably annex twelve other small rural districts. As a result, the School District of Hazelwood encompassed 78 square miles—making it seventeen square miles larger than the City of St. Louis.[36]

In 1951, voters in the Ferguson and Florissant school districts also voted to consolidate. The move allowed Ferguson to increase its bonding capacity to build new schools to accommodate rising enrollments; it also gave Florissant—then less populated than Ferguson—an improved educational program that included a public high school. The surge in homebuilding exceeded all expectations in the newly consolidated district, which had to construct fifteen schools between 1953 and 1969.[37]

The dynamics of population growth—shaped by fragmentation, annexation, and consolidation—resulted in fifteen separate incorporated municipalities in the forty-seven square miles of the Coldwater Creek watershed. Early communities and their dates of incorporation include Florissant (1786), Bridgeton (platted as Marais des Liards in 1794 and incorporated in 1843), and Ferguson (1894). They were followed by Berkeley (1937), Overland (1939), Calverton Park (1940), Sycamore Hills (1941), St. John (1945), Edmundson (1948), Kinloch (1948), St. Ann (1948), Hazelwood (1949), Breckenridge Hills (1950), Woodson Terrace (1954), and Black Jack (1970). A portion of the watershed remained unincorporated, with St. Louis County serving as the local government.[38]

These incorporations predated citizen knowledge of the radioactive contamination of Coldwater Creek. The downside of government fragmentation became clear later, when three local governments were unable to act together to protect their communities against a common threat. In 1978–79 Hazelwood, Berkeley, and Florissant took separate opposing positions when federal officials wanted to move twenty thousand tons of radioactive material through Hazelwood and Berkeley—and near Coldwater Creek, which flowed through Florissant. The next chapter explores the vulnerability of such small governmental units.

The Perfect Place

Over time, the suburbanization of north St. Louis County heightened a need for public parks and athletic fields, several of which abutted Coldwater Creek. They include St. Ferdinand Park (owned by the City of Florissant), as well as three properties operated by St. Louis County: Harold J. Evangelista Park in Black Jack, Vatterott Field in St. Ann, and Coldwater Commons Park in Florissant. Today Coldwater Creek still evokes warm childhood memories among people who came of age in those communities. Patricia Barry recalled a wooded area behind her home where the tributary meandered. It made for idyllic summers. She explained, "We'd spend all our free time outside" playing in the woods and creek, "making houses with branches and sticks and anything we could find." While Barry had long regarded the fifteen years spent in her childhood home as "good ones," she reassessed upon receiving a cancer diagnosis in adulthood and discovering the creek was contaminated.[39]

An Exception

It is important to note that the postwar population explosion in north St. Louis County was overwhelmingly White. The Coldwater Creek watershed had several Black communities, some dating back before the Civil War, when slavery was legal in Missouri. However, after World

War II, African Americans in the St. Louis area still were prevented by law, procedure, and practice from enjoying the country's overall affluence. They faced major barriers: race-restrictive deed covenants that prohibited their residence in "White" neighborhoods; "steering" by realtors; and limited access to the GI Bill and FHA loans that helped White citizens enter college, purchase homes, and start businesses.[40] These factors are further explored in chapter 6, "An Environmental Justice Watershed."

Stream Ecology

In time, there were new signs that something was wrong in the Coldwater Creek watershed. In the 1970s Herb Thies—who had farmed in the area for decades—could not raise a crop one-half mile north of SLAPS. "The land wouldn't grow anything," Thies explained. "I put in soybeans. I planted early in the spring, and after May and June, there was nothing to harvest." He added, "The outer edges worked, but the middle—it was dead dirt. It never came out right."[41]

Later, in its 2005 *Record of Decision for the North St. Louis County Sites*, the Army Corps of Engineers would report that the principal contaminants of concern in the Coldwater Creek watershed were radionuclides from residues associated with MED/AEC ore processing. Engineers also noted that the stream ecology of Coldwater Creek had been "severely affected" by "industrial and other operations" unrelated to the MED and AEC. The corps stated,

> Pollutants enter the creek in storm water from commercial and industrial facilities, residential areas, and Lambert–St. Louis International Airport. The SLAPS storm-water run-off flows into Coldwater Creek. More than a dozen facilities that are permitted under the National Pollutant Discharge Elimination System (NPDES) program discharge directly into the creek, including the Ford Motor Company, Lambert–St. Louis International Airport, and the Boeing Company. Permitted discharges include storm-water runoff and manufacturing discharges.[42]

As part of an ecological risk assessment, the corps divided the creek into three zones between SLAPS and the mouth at the Missouri River. Engineers found the existing terrestrial and aquatic habitats were "limited in extent and substantially affected by their urban surroundings." The exception was *a portion* of the third zone (between Lindbergh Boulevard and the mouth at the Missouri River) where there was a "relatively unique habitat" for animals. They included herbivorous birds (mallard ducks), piscivorous birds (kingfishers), and mammals (raccoons). The report states that future excavation of sediment from this area to protect human health should be weighed against the probability of "adverse effects," to animal and plant life, including "population reduction, habitat disruption, and sedimentation." However, it concludes that such effects would likely be temporary.[43] It is interesting to note that, in addition to "L'eau Froide," early French settlers called Coldwater Creek "Rivière Les Biches," in reference to the deer there.[44]

CHAPTER 4

Bureaucratic Blues

When the atom bomb was exploded many, many years ago, people
stood and watched it. . . . Now we know it was a dangerous thing
for them to do. They now say . . . [the field] is not dangerous.
But in 20 or 30 years from now will they still say that?

—**William Miller, mayor of Berkeley, Missouri, 1991**

When World War II ended, only one site in the St. Louis area held
uranium residues and wastes from the Manhattan Project. The
following year, the number increased to two properties: Mallinckrodt's downtown plant and the St. Louis Airport Storage Site. By
1996, Mallinckrodt's uranium residues and wastes were documented
at *eighty-two sites in St. Louis City and County alone.* Implausibly,
the radiological spread occurred on the watch of five successive federal entities charged with overseeing the contaminated material. They
were the Manhattan Engineer District (MED), Atomic Energy Commission (AEC), Nuclear Regulatory Commission (NRC), Energy Research and Development Administration (ERDA), and Department of
Energy (DOE).

Between 1946 and 1996, decision-makers in the five federal entities
could rely on one irrefutable truth: Mallinckrodt's uranium wastes and
residues posed a threat to people who had contact with them. Questions of *how dangerous* and *in what quantity* were less clear, and subject to scientific inquiry. Officials were required to make judgments on
radioactive exposure when St. Louis County was changing rapidly—
through a population explosion, new business development, fragmented political entities, growing environmental awareness, and increased citizen engagement. Unfortunately, federal officials chose the
Manhattan Project's approach during World War II. They valued ef-
ficiency, operated in secret, prioritized nuclear production above all

else, gave false assurances to the press and public, and hired contractors without closely supervising them. Today the agency officials' decisions appear scattershot and lacking in strategy. Their actions (and inaction) clearly contributed to the public health crisis in the Coldwater Creek watershed.

It is important to remember that the radiological contamination of north St. Louis County did not occur overnight. It was a slow-motion disaster that unfolded over half a century. One is therefore inclined to look back and ask, "What did governmental officials know, and when did they know it?" This chapter will answer those questions while presenting the failures of the federal agencies to protect the watershed's residents.

Atomic Energy Commission

In 1946, Congress transferred control of atomic energy from military to civilian hands by replacing the Manhattan Engineer District with the Atomic Energy Commission. In passing the Atomic Energy Act that year, Congress granted the U.S. government a virtual monopoly on atomic energy. The act's 1954 amendments gave the five-member commission three areas of responsibility: to continue nuclear weapons production, promote the private use of atomic energy for peaceful purposes, and protect public health and safety from the hazards of commercial nuclear power. J. Samuel Walker states that the three functions were "in many respects inseparable and incompatible," especially within a single agency. Over time, the AEC's attention to military and promotional duties lessened its credibility with the public on matters of regulation and safety. Protecting the public health seemed a lower priority for the agency than weapons production.[1] Nevertheless, during the 1950s the AEC became one of the largest industrial enterprises in the United States—with little congressional oversight of health and environmental issues until the 1970s.

Related to the lack of external scrutiny was the fact that the AEC supervised the nuclear weapons complex but did not actually operate it. To many, the production of nuclear weapons seemed so complicated and difficult that it required corporations to carry it out. This "government-owned, contractor-operated" system was a carryover from the Manhattan Project, for which large corporations and universities designed weapons and operated factories that created materials for the first atomic bombs. Before agreeing to do the job, corporations (like Mallinckrodt) insisted on being completely free of liability for their actions, even when they were negligent.[2]

Mallinckrodt purified uranium at its downtown location from 1942 to 1957. The AEC then moved the company's uranium operation to a new automated facility at Weldon Spring in St. Charles County, Missouri, until its closing in 1966. Mallinckrodt officials attributed the closure to a decreased demand for uranium, and the AEC shifted its uranium business to a new plant in Fernald, Ohio.[3]

Meanwhile, the AEC faced growing problems at the St. Louis Airport Storage Site. In the 1960s researchers at the commission's Mound Laboratory in Ohio found that residues and wastes at SLAPS contained the largest concentration of thorium-230 in the United States and possibly the world. (As early as 1959, experts at the Hanford Site had concluded that thorium-230 was "in a class as hazardous as plutonium.") According to Robert Alvarez, the thorium concentration at SLAPS was twenty-five thousand times greater than what would normally exist in nature. He explains, "Over a half-life of 77,500 years, thorium-230 decays to radium-226 and undergoes a substantial 'ingrowth' of alpha radioactivity. In other words, the waste becomes increasingly 'hotter.'"[4]

AEC officials handled the SLAPS radioactive waste by turning it into a commodity. From 1962 to 1964, the commission made three unsuccessful attempts to sell the "uranium-thorium source materials" at the property. In 1965 the AEC appointed a "Committee on the Disposition of St. Louis Airport Site," tasked with developing a plan to (1) remove SLAPS residues and wastes to AEC's new Weldon Spring chemical plant, (2) clean up SLAPS, and (3) dispose of the SLAPS site after cleanup. In November 1965 the committee reported that 121,050 tons of uranium residue remained at the St. Louis Airport Storage Site; furthermore, the possibility of selling the radioactive material was "remote." The group recommended that the AEC remove the material in question and, after a minor cleanup, dispose of the SLAPS site on a restricted basis.[5]

Despite the committee's pessimism about sales prospects for SLAPS's radioactive material, a buyer emerged. Contemporary Metals paid slightly more than a dollar a ton for the purchase. In 1966 its subsidiary, Continental Mining and Milling Company, began moving uranium residues and wastes from SLAPS to 9200 Latty Avenue in Hazelwood—one-half mile north of the airport site. Unfortunately, spillage occurred. According to interviews conducted in 1989 by the

St. Louis Post-Dispatch, there was waste all over the streets, and children played with truck spills "as if they were sifting for gold." Later, through a bad debt foreclosure, Commercial Discount Corporation of Chicago acquired the SLAPS process residues. In 1969—after that company went bankrupt—Cotter Corporation purchased the remaining residues and began shipping them one year later to the company's plant in Cañon City, Colorado.[6] The piles at Latty Avenue sat uncovered in order to allow the material to dry and lower shipping costs. As a result, wind and rain moved a portion of the radioactive material to the surrounding soil. Some of the wastes eventually spread to Coldwater Creek, where they could flow downstream, settle in certain locations, and remain in soils near the creek after flooding.[7]

Meanwhile, in 1967 the AEC authorized the use of SLAPS by the City of St. Louis. Then on May 15, 1973, the U.S. government and the City of St. Louis agreed to transfer the ownership of SLAPS by quitclaim deed to the St. Louis Airport Authority, an arm of city government.[8]

Cotter's drying process prior to transfer left roughly eighty-seven hundred tons of leached barium sulfate at the Latty Avenue site. (The barium had been used to recover uranium and contained uranium residue.) To dispose of it, Cotter mixed the barium sulfate with thirty-nine thousand tons of topsoil and in 1973 contracted B&K Construction Company to ship it to the West Lake Landfill in nearby Bridgeton.[9] On November 1, 1974, the AEC notified Cotter Corporation that the disposal material did not meet its standards because inspectors had concluded it was too radioactive. At issue was the agency's prohibition against altering or diluting radioactive source material to create a mixture that was no longer subject to licensing.[10] Although the AEC determined Cotter was "clearly in violation" of federal regulations, there was no enforcement action against the company. The commission mistakenly believed the radioactive deposit was buried under one hundred feet of municipal waste when, in fact, it was under just three feet of soil.[11]

Ironically, the movement of contaminants to the West Lake Landfill occurred during the 1970s—the environmental decade of U.S. history that saw the passage of landmark legislation to protect the air, water, and soil. Among the laws passed was the National Environmental Policy Act, which mandated an environmental impact statement for every proposed federal action that would affect the quality of the human environment. The decade also saw the first Earth Day, launched in 1970 by Senator Gaylord Nelson of Wisconsin as a series of teach-ins across the United States. The events' purpose was to raise aware-ness of the earth's plight and to initiate actions to save it from further degradation. Earth Day became an annual national and international event that helped spawn the modern environmental movement.[12] Fi-nally, the 1970s saw the establishment of the Environmental Protec-tion Agency (EPA), which integrated widely scattered federal research and regulatory programs that dealt with environmental pollution.[13] The significance of combining these activities into one agency—and the growing importance of the EPA—resulted in subsequent action against radioactive contamination in north St. Louis County.

Congress abolished the Atomic Energy Commission in 1974.[14] Be-fore its dissolution, however, the AEC established the Formerly Uti-lized Sites Remedial Action Program (FUSRAP) to take responsibility for the environmental remediation or control of sites where the MED or AEC had operated from the 1940s through the 1960s. The down-town Mallinckrodt plant, SLAPS, and Latty Avenue properties were included as FUSRAP sites, along with seventy-eight "vicinity proper-ties" in Hazelwood and Berkeley, Missouri.[15]

NRC and ERDA

Congress replaced the AEC with two new agencies: the Energy Re-search and Development Administration (ERDA) and the Nuclear Regulatory Commission (NRC). The former handled energy research and development, nuclear weapons, and naval reactor programs; the latter regulated civilian uses of atomic energy (mainly the commercial nuclear power industry). In 1977 ERDA merged with the Federal En-ergy Administration to form the Department of Energy (DOE), while the NRC remained a separate entity.[16] Like the AEC, it failed to act against Cotter at the West Lake Landfill. The NRC had the legal right to ask that the radioactive waste be retrieved and placed in suitable storage; however, it "let the Cotter company off the hook" by terminat-ing its license to possess the material.[17]

The 1996 report of a DOE-appointed citizens' group explains the significance of the bureaucratic reordering that occurred in the mid- to late 1970s. It contends that splitting the AEC into two agencies meant those branches would be "self-regulating [and therefore not held ac-countable] until the passage of the Superfund Amendments and Re-authorization Act in 1986, which . . . [would] bring the federal facility cleanups under EPA oversight."[18]

In 1977, E. Dean Jarboe purchased 3.5 acres in the 9000 block of Latty Avenue to build a headquarters for his company, Futura Coatings. Three days after the sale was final, federal officials informed Jarboe that he could not use his newly acquired property because it was contaminated. In 1980 Jarboe bought an adjacent 7 acres to serve as an interim storage site for the contaminated and demolished building rubble cleared from his 3.5-acre lot. Although he planned to expand his operations on this land after the federal government removed the radioactive material, the debris stayed in place until 1998, following an action memorandum.[19] The consolidated waste on Latty Avenue became known as the Hazelwood Interim Storage Site (HISS).[20]

As Jarboe awaited the removal of radioactive waste from his property, the NRC issued a June 1988 report describing the extent of contamination at West Lake Landfill in Bridgeton. The document detailed the presence of radium-236, uranium-238, uranium-234, and thorium-230. Based on this information—and the knowledge that radioactive material becomes more dangerous as it decays—the NRC staff concluded that radiological hazards would increase at West Lake in the future. They urged that measures be taken to establish permanent control of the site. However, the NRC did not indicate it would take any further action at the West Lake Landfill.[21] (Only six years prior, the U.S. Nuclear Regulatory Commission calculated that highly mobile radon gas generated from the ingrowth of radium would increase fivefold in one hundred years and nearly double that amount in two hundred years.)[22]

Department of Energy

In 1982 the Department of Energy was pursuing its own course of action by favoring the disposal of SLAPS and Latty wastes at Weldon Spring in nearby St. Charles County, Missouri. It was the former site of the U.S. Army Ordnance Works during World War II and a Mallinckrodt facility for uranium refining from 1957 to 1966. An estimated two thousand people attended a hearing in St. Charles County in opposition to DOE's plan. U.S. senator Thomas Eagleton of Missouri introduced a bill for DOE to *reacquire* the St. Louis Airport Storage Site from the city and study options for disposing of SLAPS and Latty wastes. Although Eagleton's legislative proposal became law in 1984, the City of St. Louis did not immediately transfer the property back to the Department of Energy. As St. Louis officials resisted, the DOE reviewed design options for disposing of the SLAPS and Latty Avenue residues and wastes *at the St. Louis Airport Site*. In 1988 the city board of aldermen's Special Committee on Radioactive Waste issued a report urging the Missouri congressional delegation to direct DOE to find an environmentally sound disposal site for the St. Louis wastes—far away from a major population center. In a nonbinding referendum, 85.6 percent of St. Louis County voters opposed constructing a radioactive waste bunker at SLAPS, while 80.77 percent of St. Louis City voters also registered their disapproval of such a bunker at SLAPS or any other location in the city.[23]

Meanwhile, beginning in the early 1980s, erosion from the St. Louis Airport Storage Site was reported to be slowly entering Coldwater Creek. In 1985 the DOE attempted to curb the erosion by constructing a gabion wall on the creek bank.[24] Subsequent tests revealed the presence of thorium-230 extending from the SLAPS site for several miles downstream.[25] The following year, the Army Corps of Engineers documented the DOE's knowledge of and response to radioactive contamination of the creek. In a report titled *Coldwater Creek, Missouri: Feasibility Report and Environmental Impact Statement*, the corps noted the following:

> Radioactive materials are located at two storage sites adjacent to Coldwater Creek downstream from Lambert Airport [SLAPS and HISS]. Some radioactive material has entered Coldwater Creek. In 1986 and 1987 the Department of Energy (DOE) is conducting studies to determine if there are significant radioactive materials in the Coldwater Creek channel. If significant radioactive contamination is found in the channel, the DOE will develop plans for remedial action. Any DOE project would not be initiated until the 1989 or 1990 time frame, but would be expected to be completed before construction of the proposed Corps of Engineers flood control and recreation project.[26]

The corps also issued a cautionary note, stating, "During the past forty years the Coldwater Creek watershed has changed from a predominately rural area to a highly urban area with many concrete channels on tributary streams. Flood levels from a given storm will be higher on Coldwater Creek now as a result of this development."[27]

In 1992, the Federal Facilities Compliance Act extended EPA's and the states' authority to impose sanctions against hazardous wastes at federal facilities. That same year EPA initiated Superfund enforcement action against potentially responsible parties at West Lake Landfill. Included among this group were the U.S. Department of Energy and the Cotter Corporation. DOE claimed it had no responsibility for wastes at the landfill because they were sold for their commercial value and not as a mechanism for disposal. However, the EPA noted that AEC's "Instructions to Bidders" (in the attempted sale of Latty residues) indicated the barium sulfate was waste. Accordingly, EPA continued to hold the DOE at least partially responsible for the debacle at the landfill.[28] In 1993, EPA issued a consent order against the potentially responsible parties at West Lake requiring them to conduct a remedial investigation / feasibility study on the site.[29]

Appointed by the Department of Energy, the St. Louis Site Remediation Task Force formed in September 1994 as a "broadly-based representative body" that included people from DOE, EPA, U.S. congressional offices, and city and county commissions—as well as vicinity property owners and representatives from Mallinckrodt and the utility companies. The group was charged with identifying and evaluating remedial-action alternatives for the St. Louis FUSRAP sites and the West Lake Landfill. The task force was also responsible for petitioning the DOE to undertake a cleanup strategy that was "environmentally acceptable and responsive to public health and safety concerns."[30] Members approached their work with an understanding that remediation would be a complicated task magnified by a unique set of factors, which they described as follows:

- Contaminated sites are located in a densely populated metropolitan area of 2.4 million people
- Industrial, residential, and recreational activities have occurred and continue to occur in and around the contaminated sites
- Evidence indicates that extensive migration of radioactive contaminants by air, surface water and ground water transport has occurred and current evaluations suggest ongoing migration
- Contaminated properties, primarily single family residential and commercial developments, are in an urban flood plain
- Coldwater Creek, within a 47-square-mile urban watershed, contains radioactive sediment
- The Post Maquoketa Aquifer, which lies beneath the St. Louis Airport Site (SLAPS) and extends north through St. Louis County under many of the radiologically contaminated properties, is the only bedrock aquifer yielding potable water in northern St. Louis County

- NO community acceptance exists for a permanent repository at SLAPS or in St. Louis[31]

The task force evaluated possible radioactive waste disposal sites and concluded that SLAPS was not an appropriate one; neither were other disposal sites in the St. Louis area, for a variety of reasons. Reiterating some previously described unique factors and adding others, members considered a range of elements, including location in a floodplain or in a densely populated area, proximity to groundwater, unsuitability of geologic strata, proximity to heavily traveled roads, possibility of contaminant migration, presence of uncontrolled accessibility, potential for negative impact on real estate values and economic development, and absence of an appropriate disposal facility.[32] Although the task force chose not to recommend a suitable site for FUSRAP's waste, the group felt such materials should be housed in a licensed commercial site. Members also recommended that West Lake Landfill be fully encapsulated.[33]

In making cleanup recommendations, the St. Louis Site Remediation Task Force sought a standard of radiation exposure that did not cause harm. The DOE, in turn, provided its standard for land deemed suitable for unrestricted use.[34] However, task force members noted that since 1957 (when the federal government had begun setting allowable radiation exposure standards), the data suggested an ongoing need to make the standards progressively more stringent. So, while using DOE's formula, the group recommended a conservative approach in applying it.[35]

In its conclusion, the task force report singles out SLAPS as the first cleanup priority for the DOE. Next are the Ballfields in Berkeley, North City and North St. Louis County Properties and Haul Roads, HISS / Futura Coatings, and Coldwater Creek (not necessarily in that order). The report also cites the St. Louis Downtown Site, West Lake Landfill, and City Levee (Riverfront Trail) for remediation.[36]

With respect to Coldwater Creek, one task force working group reported, "Insufficient data exist to make any judgments regarding the long-term health and environmental effects of the contamination." The working group recommended a creek cleanup using the most stringent standards in order to minimize health and safety risks and stop ongoing contamination of the creek and downstream properties.[37] In its report, the task force also included the findings of a panel of geologists and hydrologists who stated that the SLAPS contamination of Coldwater Creek and the groundwater was "not acute at this time." However, the panel added, "[The pollution] does present a chronic problem to environmental quality along . . . [the creek] and should be corrected."[38] Specifically, Coldwater Creek sediments contained radionuclides that extended downstream from the SLAPS site. The panel further reported,

> Radioactive contamination soil at SLAPS has been characterized and extends to a depth of about 18 feet, with the majority of contamination between 4 and 8 feet below land surface (bls). Levels of uranium-238, radium-226, thorium-230, and thorium-232 in the samples from these depths exceed background levels. Results of groundwater analyses in some monitoring wells, stormwater, and Coldwater Creek sediment also indicate elevated uranium levels. However, measured levels of radionuclides in surface water from Coldwater Creek were consistent with background levels and lower than proposed Department of Energy Guidelines.[39]

In keeping with one task force recommendation, the Army Corps of Engineers began the cleanup of St. Louis Airport and Ballfield Sites in March 1998. (Only one year earlier, Congress had transferred the nation's nuclear waste cleanup to the corps from the Department of Energy.) The airport and ballfield project would end with a formal closing ceremony in 2007—the culmination of moving six hundred thousand cubic yards of radiologically contaminated material. In 1999 the corps also constructed a HISS / Latty Avenue rail spur to facilitate removing radioactive wastes from various properties on that street, a task completed between 2010 and 2013.[40] These actions were consistent with a final Record of Decision issued in 2005 by the Army Corps of Engineers (fifty-nine years after radioactive material had first arrived at SLAPS). The report outlines the selected remedy for cleaning up FUSRAP sites in north St. Louis County.[41]

State and Local Responses

In some cases, state and local governments weighed in on the radioactive waste in the Coldwater Creek watershed. In 1976, the Missouri Department of Natural Resources (MDNR) informed the National Regulatory Commission that articles on uranium dumping at the West Lake Landfill had appeared in the *St. Louis Post-Dispatch*. Created in 1974, the MDNR asked the NRC to investigate and reassess the West

Lake disposal. After the request, NRC followed up with both Cotter Corporation and West Lake.[42]

In 1979, the Florissant City Council unanimously approved a resolution opposing the pending removal of twenty thousand tons of "low-level radioactive dirt" from Latty Avenue in Hazelwood. The dirt would have been carried through Berkeley for storage at SLAPS. Council members expressed concern for their constituents since both the Latty and airport sites were adjacent to Coldwater Creek, which flowed through Florissant. The council wanted the contaminated material to remain in Hazelwood until an alternate storage site away from a heavily populated area could be found. Florissant's position differed from that of Hazelwood (which wanted the dirt removed) and Berkeley (which wanted it moved to the airport under "strict safety restrictions").[43]

In 1989, the EPA placed both the SLAPS and Latty properties on the Superfund National Priorities List. The list serves as a guide in determining which sites in the United States are so contaminated as to warrant further investigation and significant cleanup. In the face of inaction from both the DOE and NRC, the Missouri Department of Natural Resources asked the EPA to include West Lake Landfill as a Superfund site. This was accomplished in 1990.[44]

Increasingly, citizens were expressing dissatisfaction with the Department of Energy and its handling of radioactive material in their north St. Louis County neighborhoods. In 1989 the *St. Louis Post-Dispatch* ran a seven-part series on the bomb's legacy in the city. Two years later, the *Los Angeles Times* also covered the controversy around radioactive waste in the St. Louis suburbs. *Times* writer Eric Harrison characterized DOE's handling of radioactive waste by reporting, "The Department of Energy, which says the material does not pose a health hazard, at one time favored permanently storing it here in a bunker to be built near Lambert International Airport. Now the DOE is conducting a long-range study to determine what ultimately should be done." Harrison noted that citizens and local officials who were upset by the radioactive waste had mounted a campaign for its removal.[45]

Harrison interviewed William Miller, mayor of Berkeley, which bordered the St. Louis Airport Storage Site. He stated that for more than twenty years Berkeley had operated a ballfield across from SLAPS. Although DOE gave assurances that the field was safe and could only cause harm if people ate the dirt, Miller had his doubts. He noted that some Berkeley residents lived within a half mile of the radioactive site, and a quarter of a million people lived within a three-mile radius of it. Expressing concern that a DOE bunker could store radioactive waste from other parts of the country, Miller observed, "If this becomes a federal site, there's not a state or local government that can do anything about it.[46]

On one occasion a local government worked with federal officials to address radioactive contamination, with unsatisfactory results. In 1986, the Department of Energy secured Hazelwood's permission to temporarily place 3,700 yards of contaminated soil from a sewer installation along the Hazelwood-Berkeley boundary. Time passed, and the dirt remained. Eventually the Federal Emergency Management Agency (FEMA) ordered Hazelwood to remove the soil, contending it had altered the floodplain. The agency voiced concern that the dirt might spread radioactive contamination downstream along Coldwater Creek. (There was a one-in-one-hundred chance, in any given year, that the contaminated soil would be inundated by floodwaters.) Absent a resolution of the matter, Hazelwood would lose flood insurance, along with federal loans and grants. Moreover, without flood insurance, lending institutions would refuse to make loans to homeowners and businesses located on a floodplain. When Hazelwood was unable to find another site to accept the soil, local and federal officials collaborated to achieve a creative "solution": redrawing the map so that the side of the creek containing the dirt was out of the floodway.[47]

Who Knew?

Until 1978, few people had knowledge of the nuclear waste in the St. Louis area. Implausibly, the news spread after a chance meeting of a nun and a scientist on a flight from New York to St. Louis. Nuclear-disarmament activist Sister Mary Ann McGivern was seated next to Cornell University physicist Robert Pohl, who was traveling to testify about uranium mining on Native American lands in the West. As they chatted over dinner, Pohl asked McGivern whether she was fighting for the cleanup of nuclear waste in St. Louis.[48]

"What waste?" McGivern asked.[49]

Pohl recalled the location as "Laddie Avenue," having recently read about it in a report. McGivern followed up and contacted St. Louis County resident Kay Drey to lead the initiative. She was a logical choice, having blocked Union Electric's plan to build a second nuclear reactor in Callaway County, Missouri. The *St. Louis Post-Dispatch* reported that, because of the "uproar" caused by the environmentalists, the contaminated material stayed at Latty Avenue.[50]

Beginning in 1979, Drey led a "pitched battle" against the NRC after it made public plans to transport contaminated material from Latty Avenue and combine it with other radioactive waste at SLAPS. Once the Latty Avenue waste was deposited there, officials planned to pave over part of the SLAPS property and turn it into a driver training course for local police departments. Drey "bombarded" political leaders with letters, petitions, and protests and sparred with NRC official William Crow during municipal meetings. She also produced a federal report that radioactive waste was already leaching from SLAPS into Coldwater Creek.[51]

Drey's activism and Senator Eagleton's legislative proposal sparked the formation of the ad hoc "Citizens for a Safe North County" organization. In a flyer for general distribution, the group warned what would happen unless residents protested: "The federal government will turn a temporary radioactive waste dump into a permanent one." The flyer linked SLAPS deposits to the "manufacture of atomic bombs" and cautioned, "[Radioactive wastes] are eroding into roadside ditches along McDonnell Blvd. and on into Coldwater Creek . . . which periodically floods in Hazelwood, Florissant, and northeast St. Louis County."[52]

Newspapers also played a key role in publicizing the issue. As early as September 19, 1978, the *St. Louis Globe-Democrat* reported on misgivings expressed by Berkeley City Council members when meeting with NRC representative William Crow. He told the council and some thirty residents in attendance that twenty thousand tons of (low-level) radioactive dirt must be moved from a privately owned, noncontrolled site to a location where it could be publicly controlled. Beginning October 1, tarpaulin-covered trucks would transfer the dirt from 9200 Latty Avenue in Hazelwood, traveling through Berkeley to SLAPS. Crow assured the council and audience that safety personnel would check the trucks for radioactivity at both sites, perhaps watering down the dirt to prevent blowing. Environmental monitoring stations would also check the air quality during transport, and new wells at SLAPS would allow groundwater to be monitored for possible contamination. Crow confirmed that a layer of dirt and asphalt would eventually cover the transferred material so that the site could become a police driving course.

Citizens in attendance expressed concern about wind blowing the dirt, spillage during transport, and long-term effects of radioactive contamination in the Berkeley community. In response to one resident, Crow admitted that he could not guarantee the NRC would immediately find spillage. Another citizen challenged Crow's contention that the material was of minimal danger, noting, "Scientific studies show there is no safe level of exposure."[53]

One year later, the *St. Louis Globe-Democrat* reported that "thousands of tons of radioactive dirt spread among four St. Louis area storage sites" were the byproducts of Mallinckrodt uranium processing for the first atomic bomb. The newspaper detailed the transport of "contaminated dirt" from Mallinckrodt to the northeast corner of the municipal airport, the spreading of SLAPS contamination to Latty Avenue, the further contamination of West Lake Landfill, and the relocation of "4,300 truckloads of contaminated dirt" to a new Mallinckrodt plant at Weldon Spring in St. Charles County.[54]

The *Riverfront Times*, founded in 1977 as St. Louis' alternative news weekly, ran an August 1982 cover story with the heading "Danger in Our Own Back Yard: Hazardous Waste from World War II Bombs Live on in St. Louis." In the same issue, Ray Hartmann wrote a column on hazardous waste for the same issue in which he observed, "One of the first lessons any of us learn in life comes in the form of the simple directive, 'Clean up your own mess.'"[55]

In 1983 the *Globe-Democrat* closed, leaving the *Post-Dispatch* as St. Louis's only metropolitan daily newspaper. Importantly, the *Post*'s 1989 series on the radiological contamination of the St. Louis area discusses the World War II association with the Manhattan Project, the radioactive waste storage at SLAPS, the health of uranium workers, the contamination of Weldon Spring, the efforts of local activists, and the positions of government officials on various related issues. Written by Carolyn Bower, Lou Rose, and Theresa Tighe, the series originated with Rose himself. He clipped news stories of infant deaths in St. Charles County and wondered whether there was a connection to the shuttered building at Weldon Spring where Mallinckrodt had operated a uranium processing plant.[56]

What Did They Know, and When Did They Know It?

Information uncovered by activists and the press begs the question of whether scientists and/or government officials knew the real risks of uranium refining and didn't share them with the public. The history of radiation protection suggests that scientists have long known radiation exposure adversely affects human beings; however, their understanding was still somewhat imprecise, even in 1996. In *Permissible Dose: A History of Radiation Protection in the Twentieth Century*, J. Samuel Walker explains that as early as the 1920s, scientists worked to define a level of radiation exposure that did not produce observable injury. In 1934 both the U.S. and international radiation protection committees felt they had enough information to recommend a quantifiable "tolerance dose" of external radiation—that is, an exposure level below which one could generally assume to be safe. Although their work was based on "imperfect knowledge and unproven assumptions," committee members felt that practical guidelines would reduce injuries to radiation workers. In 1941 the American panel went further to recommend tolerance doses for the principal sources of hazard from internally deposited radiation, radium, and its decay product, the radioactive gas radon.[57]

In 1948, two years after uranium waste had been transported to SLAPS, the Atomic Energy Commission issued a warning, stating, "The ultimate disposal of contaminated waste—subsurface, surface, or airborne—needs much more thorough study. Even the simplest of such data. . . . are almost wholly lacking." The AEC added, "The disposal of contaminated waste in present quantities and by present methods (in tanks or burial grounds or at sea) if continued for decades, *presents the gravest of problems.*"[58]

The bombs at Hiroshima and Nagasaki had made radiation safety a more difficult task because nuclear fission created many nuclear isotopes that didn't exist in nature. Not only was less known about them, but the number of people exposed to radiation was likely to increase. To gain more information (especially about plutonium, which was in the Nagasaki bomb), government agencies engaged in some instances of appalling behavior.[59]

From 1945 to 1947, eighteen patients received plutonium injections (all but one without their consent) under the auspices of the Manhattan Project and the Atomic Energy Commission. The experiments took place at several sites, including Oak Ridge Hospital, the University of Rochester, the University of Chicago, and the University of California. The purpose was to learn about the movement of plutonium through the body—information later used in determining permissible doses of the element. Most patients did not know that plutonium was injected into their bodies or even that they were subjects of an experiment. (Research standards of the day did not require that a subject be informed of the nature of the experiment or give documented, informed consent.) Officials were convinced that the doses of plutonium were too small to produce long-term consequences (which were more problematic than short-term ones). Therefore, they chose patients who were not expected to live long. Surprisingly, several of them lived for many years.[60]

In 1946—and again between 1950 and 1953—the AEC, National Institutes of Health, and Quaker Oats Company funded research conducted by scientists at MIT. They fed oatmeal containing radioactive trace elements of iron or calcium to more than one hundred boys at the Walter E. Fernald State School in Waltham, Massachusetts. Many of the students, ages twelve through seventeen, were incorrectly labeled as mentally retarded. The tests provided information on how the body absorbs iron, calcium, and other minerals in food. However, as in the plutonium injections, the experiments were not designed to provide any health benefit to the subject. Although the school asked parents to sign a consent form, it described the purpose of the experiment in vague terms and failed to mention the ingestion of radioactive materials. To encourage participation, the boys were told they would be part of a "science club" and would receive special treatment, such as trips to baseball games and extra portions of milk at meals.[61]

Also in 1946, the National Committee on Radiation Protection (NCRP) reassessed its position on radiation exposure levels. It replaced "tolerance dose" with "maximum permissible dose" to better convey the idea that no amount of radiation was certifiably safe. NCRP defined permissible dose as that which "in the light of present knowledge, is not expected to cause appreciable bodily injury to a person at any time during his lifetime." In formulating the definition, NCRP acknowledged the possibility of suffering harmful consequences from radiation in *amounts below* the permissible limits. However, the committee's understanding of permissible dose was based on the belief that "the probability of the occurrence of such injuries must be so low that the risk would be readily acceptable to the average individual." In response to the expected growth of atomic energy programs and a substantial increase in the number of individuals involved, the NCRP also revised its recommendation on radiation protection. It reduced the permissible dose for whole-body exposure from external sources to 50 percent of the 1934 level.[62]

The important point here is that no exposure to radiation is entirely safe, and any amount involves some risk. Since uranium is naturally radioactive, with an unstable nucleus, it is in a constant state of decay as it seeks a more stable arrangement. It becomes more dangerous as disintegration occurs. As noted by the St. Louis Site Remediation Task Force in 1996, research data continually suggested a need for *more stringent* standards. Therefore, government officials were dealing with a moving target. It was difficult to find a balance between personal safety and the increasing demand for energy.

Under the Superfund Amendments and Reauthorization Act of 1986, the EPA refined its risk calculation for assessing the hazardous wastes. Under the new guidelines, statisticians started with the "background" rate of cancer—about 2,500 to 3,000 cancers per 10,000 people in the United States. Anything over the background rate was considered "excess" and subject to further evaluation.[63]

A Letter

Despite the difficulties of navigating this balance, federal officials clearly knew from testing *in the mid- to late 1970s* that radioactive material was migrating into Coldwater Creek. The Army Corps of Engineers states that from 1976 through 1978, Oak Ridge National Laboratory conducted a "radiological investigation of SLAPS." It found elevated concentrations of uranium-238 and radium-226 in drainage ditches north and south of McDonnell Boulevard.[64] Even more telling is a letter to U.S. representative Robert A. Young of Missouri from Ruth C. Clusen, assistant secretary for environment at the U.S. Department of Energy. In response to Young's inquiries, Clusen stated the DOE had conducted a radiological site survey in November 1976 that confirmed "some erosion of the landfill area" into Coldwater Creek.[65] She explained that sediment samples from the creek bed near the railroad tracks showed concentrations of radium-226 "10 times the natural background levels." Farther downstream (near where Brown Road / McDonnell Boulevard crosses the creek), the radium-226 concentrations were "about equal to natural background levels." Water samples from another location (near where the railroad crosses Coldwater Creek) found radium-226 concentrations of one-and-one-half times the natural background levels. Yet six hundred meters farther downstream, the radium content in water was "in the range of natural background levels." Clusen concluded, "It appears that there is some small amount of radioactivity migrating into Coldwater Creek, but the concentration becomes so diluted a few hundred meters downstream that it cannot be distinguished from naturally occurring radioactivity."[66]

Assistant Secretary Clusen reported results from additional testing of creek water and creek bed sediment in 1978:

> Both water and sediment samples showed no levels of uranium-238 or lead-210 above natural background levels in that area of Missouri. This data indicates that without additional precautions, it is likely that erosion along portions of Coldwater Creek may continue and some radioactive material may be transported to Coldwater Creek via drainage from the site. However, during such times of heavy rain and the resultant run-off from the site, it also appears that dilution of this radioactivity in Coldwater Creek occurs very rapidly to the point where the radioactivity cannot be distinguished from natural radioactivity.[67]

In fact, radium-226, with a half-life of 1,600 years, is a decay product of uranium-238. Contaminated stream sediment can settle to the bottom of a tributary and be picked up and suspended again by water that moves it farther downstream.[68]

Another Sink

In 1995, the Department of Energy began shipping radioactive waste from the St. Louis area to a Utah disposal facility run by Envirocare. When the Army Corps of Engineers assumed responsibility for the nation's nuclear waste cleanup, it proceeded in a timelier manner than the DOE. In 1998 the corps shipped to Utah twenty-four thousand cubic yards of wastes and residues from SLAPS and the downtown and Latty Avenue sites. Then in 1999–2000 the corps moved fifty thousand cubic yards of additional residues and wastes to the same destination while continuing cleanup at the St. Louis Airport Storage Site.[69]

In 2000, the *St. Louis Post-Dispatch* heralded the hazardous waste cleanup activity after "virtually no effort to control contamination for 20 years." Stating that officials had misinformed the newspaper in 1946 about the radioactive material at SLAPS, writer Virginia Baldwin Gilbert welcomed the government's more transparent approach. However, she warned readers that thorium-230 (a byproduct in the purification of natural uranium ore into uranium-235) is considered as toxic as plutonium or bomb-grade uranium. Gilbert observed, "Authorities have no way of measuring how many particles contaminated by the substance may have gotten into the lungs and guts of residents of Florissant, Hazelwood and other north St. Louis County communities."[70]

Unfortunately, it would be many years before the Army Corps of

Engineers tested for radiological contamination in the portions of Coldwater Creek north of Interstate 270, including Florissant, some parts of Hazelwood, and other north St. Louis County communities.

Looking Back

Government officials were clearly aware of the hazards of radioactive exposure well before fully sharing the information with the public. Moreover, waste cleanup was not a priority for the responsible federal agencies. DOE correspondence suggests this lack of interest in cleanup was not confined to the St. Louis area.

In 1980 the Department of Energy asked the Oak Ridge National Laboratory to conduct a preliminary radiological survey of the former Staten Island warehouse that had initially "stored" the Congolese ore before it was shipped to Mallinckrodt for refining. The depository was found to be contaminated with uranium-238 and radium-226. In response, the DOE sought an opinion from the Aerospace Corporation on whether the federal government had authority to pursue remedial action at the warehouse under FUSRAP. (The nonprofit corporation was founded in 1960 with strong ties to the military and defense industry.)[71]

In a 1985 letter to Arthur Whitman of the DOE, Edmund A. Vierzba of the Aerospace Corporation reported that the department lacked authority to take remedial action because (1) "the Manhattan District purchased only the majority of the uranium content of the ore," and (2) the U.S. government "did not have custody of the ore on the site."[72]

Susan Williams of the University of London suggests otherwise. She notes that FDR's Uranium Committee thought it was a matter of "urgency" to obtain Congolese pitchblende, especially after hearing that German scientists had an "extensive" program in uranium research. Williams adds that, even before the United States entered World War II, the committee had encouraged Edgar Sengier to move Union Minière's mined supplies of Congolese uranium for storage in New York. General Groves sent his second-in-command, Kenneth Nichols, to negotiate an agreement for America to obtain a right of first refusal on the sale of all available Congolese ore stored in a warehouse belonging to Archer Daniels Midland Company. Upon arrival, "the ore . . . was transferred immediately to the Corps of Engineers,"[73] according to Williams. When a Canadian mining company attempted to import five hundred tons of the ore from the warehouse, the U.S. Army arranged through the State Department to place export controls on it. The United States then negotiated the amount that would be made available to Canadians while continuing to import more Congolese uranium into America.[74]

Following Vierzba's opinion, the Staten Island warehouse was not included among the FUSRAP properties. In consulting with the Aerospace Corporation on matters of DOE responsibility, the department seemingly got the answer it wanted.

PART THREE

The People

CHAPTER 5

The Advocates

Human history becomes more and more a race
between education and catastrophe.

—H. G. Wells

One year after the Army Corps of Engineers celebrated the remediation of SLAPS and the Berkeley Ballfields, McCluer North High School in Florissant held a twenty-year reunion. Like many in the class of 1988, Janell Rodden Wright joined Facebook to gain information about the event. Afterward, she continued social media contact with her class.

Wright quickly knew that something was wrong. Many classmates and their children were suffering from cancers or other unusual diseases. The number of deaths in this group seemed so high that she could not attribute it to bad luck. Wright had lived in the same house for the first twenty-seven years of her life and was an accountant by training. She remained well connected to the community where she had grown up.[1] "Knowing statistics, knowing the conservative lifestyle these people had, I knew something was odd," Wright said. "In a six-house radius around my home, I knew four people with brain cancer, including a child. I looked up old neighbors on Facebook and asked them if they were OK, and they were not OK," she added. Wright continued, "Someone had lymphoma, someone had autoimmune disease, someone had lupus, some were having babies with birth defects."[2]

The numbers concerned Wright so much that her husband suggested she keep a list of ill or deceased friends and community members. By 2011, the list had grown to 274 people. "We were going to funerals frequently or visiting sick friends," Wright recalled. "It was heartbreaking." Rumors circulated that there was something in the creek running by their neighborhoods that was killing people.[3]

The subsequent discovery of Wright and her McCluer North classmates resulted in a new wave of citizen engagement to counter the nuclear contamination of the Coldwater Creek watershed. This initiative drew on earlier efforts of activists, journalists, and citizens and culminated in the federal government's acknowledgment that neighborhoods in north St. Louis County harbored radionuclides.

Just the Facts

Around the same time, McCluer North graduate Jeff Armstead created a Facebook group called "Coldwater Creek—Just the Facts Please" as a place where former residents could report their illnesses. Wright joined the group of approximately twenty people. So did former classmates Diane Whitmore Schanzenbach (an economist) and Kim Thone Visintine (an engineer), whose son Zach was diagnosed with an aggressive brain cancer at birth. After Zach died in 2006, Visintine went back to school to become a registered nurse.[4]

There was an overwhelming response to the Facebook group, which numbered 20,500 members by 2019.[5] When a group of eight to ten core members investigated the possibility of a cancer cluster in north St. Louis County, they discovered that Coldwater Creek had been contaminated with uranium processing waste. They also learned about the history of dumping radioactive material, years of neglect, and health impacts of radioactive contamination. "I think we didn't sleep for two nights," Wright recalled. "It became clear to us, from the science and medical standpoint, that what we were intuitively suspecting made sense. The puzzle fit together. We had a huge environmental issue beyond the worst imagined."[6] The group was buoyed by the addition of Angela Helbling, who played a key role in locating documents related to their inquiry.[7]

By 2013—after many illnesses had been reported to Facebook—Wright and Schanzenbach developed an online health survey for residents. The two women tracked diseases in the neighborhood using "'door-step' epidemiology," according to Robert Alvarez. Within two years they had documented seven hundred cases of cancers and immune system diseases within a four-square-mile area. Some of the illnesses (such as cancer of the appendix) appeared with relative fre-

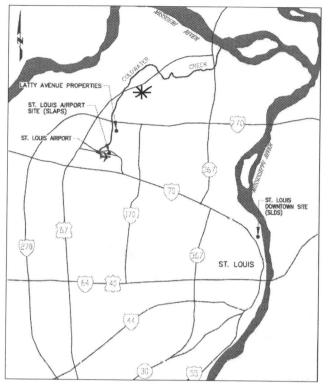

This map shows the St. Louis Downtown Site (SLDS), the St.
Louis Airport Storage Site (SLAPS), and the Latty Avenue
properties, which received contaminated material from
SLAPS beginning in 1966. The asterisk locates Florissant,
the hometown of many Coldwater Creek Facts members.
Along with other communities north of Interstate 270,
Florissant remained untested for radionuclides until 2014.
U.S. ARMY CORPS OF ENGINEERS

quency among survey respondents despite having an extremely rare
chance of occurrence.[8] Wright and Schanzenbach found higher-than-
average levels of leukemia, rare brain tumors, and breast and colon
cancers—all associated with nuclear radiation exposure, according to
the Centers for Disease Control (CDC) and the EPA. At the time, however, remediation efforts for radiological contamination were only occurring south of Interstate 270—in areas that did not include Florissant.[9] With this new information, the Coldwater Creek Facts group
sought to reach out to officials and area agencies and *educate* citizens
on the health and environmental issues in their community.

Overcoming Challenges

It was a tall order on both counts. Officials were often unresponsive, according to Wright, who in 2015 described the experience of "literally calling an agency 30 times and not having them return your phone call."[10] Moreover, the EPA promoted a tepid view of environmental education that differed from widely accepted practices of professional educators. This view was also at variance with the approach of Coldwater Creek Facts members.

In passing the Environmental Education Act in 1990, Congress tasked the EPA with increasing environmental literacy in the United States. Toward that end, the act established the Office of Environmental Education to implement a program for elementary and secondary students and "senior Americans."[11] However, the EPA asserted that, unlike government-funded health education programs (for example, CDC smoking cessation initiatives), environmental education *does not articulate a particular viewpoint or advocate a course of action*. Instead, it teaches individuals how to weigh various sides of an issue through critical thinking, problem-solving, and decision-making skills. In adhering to this position, EPA drew a sharp distinction between "Environmental Education" and "Environmental Information." The latter, according to EPA, presents people with facts and sometimes takes an advocacy stance—and therefore *is not educative*.[12]

This approach has several problems. In cases where lives are at risk, it is unthinkable that an educator would adopt a neutral view. Moreover, in an environmental crisis shrouded in secrecy, acquiring information is essential to becoming educated. It is impossible to think critically, solve problems, or make decisions without factual knowledge. Finally, while professional educators readily teach the higher-order thinking skills that EPA associates with environmental education, they also view the acquisition of information as an important part of the learning process. From the work of educational psychologist Benjamin Bloom in 1956 to the present time, colleges of education have emphasized six major categories of learning—knowledge, comprehension, application, analysis, synthesis, and evaluation. Although Bloom's taxonomy was revised in 2001, its basic framework holds. It also informs the development of teaching objectives recommended by the North American Association for Environmental Education. For Bloom and the NAAEE, each category of learning has an educative value—including the acquisition of information.[13] To this day, EPA continues to separate environmental information from environmental education. By contrast, members of the Coldwater Creek Facts group chose the dissemination of information as just one means of educating officials and the public about an environmental crisis. While engaging in critical thinking, the organization also took an advocacy stance.

Visintine cited one key point the group tried to emphasize with health researchers. She drew on her son's experience to explain that chronic exposure to low-level ionizing radiation can take decades to manifest in exposure victims or their children. Visintine stated, "I lived 27 years in the area and moved to a different zip code, 10 miles away, when my son was born. So he was never captured in the disease registry data" that states are required by law to keep, but only by zip code. She added, "That is why we started the health survey, to capture where people grew up here. . . . This was the only way we could tie things back to the creek."[14]

From the outset, the group pressed state and federal health authorities to begin investigations to determine whether a link existed between radiation and neighborhood illnesses. However, only after Wright had shown officials her initial anecdotal list of ill people—and had spoken to a local television station about her suspicions—did the Missouri Department of Health and Senior Services respond by conducting its own survey.[15] Released in 2013, the study selected four zip codes adjacent to Coldwater Creek during the period from 1996 to 2004. Researchers compared the observed number of cancer cases in each zip code against the expected number. After analyzing the data, they concluded that an increased cancer risk from radiation exposure was "unlikely." The report explains that cancer is common; it also recommends healthy eating, regular physical activity, and tobacco control as best means of prevention.[16]

Upon reading the report, Wright and Schanzenbach criticized the study's methodology, noting it did not account for people who lived near Coldwater Creek from the 1960s to the 1970s and later moved away.[17] (In fact, the zip code where the two women grew up had experienced a turnover rate of over 75 percent in the previous two decades.) In one article, Schanzenbach likened the Missouri Department of Health study to the "old joke" about the man who searched for his keys under a streetlight after losing them in a dark alley. When a passerby asked him why he wasn't looking in the alley, the man replied,

"Because the light is better over here." Schanzenbach added that the department's study excluded some important zip codes from its analysis, as well as the many cancer cases diagnosed since 2004.[18]

In September 2014 the Department of Health and Senior Services revised its study to account for the population shift. It found higher-than-normal rates of some cancers in the population, including leukemia and breast and colon cancers. The new study also revealed higher rates of childhood brain cancer in some areas around the creek. These findings aligned with Wright and Schanzenbach's survey results. However, a more comprehensive investigation was needed to directly link these diseases to radioactive contamination exposure. Accordingly, the St. Louis County Department of Public Health asked the Centers for Disease Control to conduct further studies in the area. The health department and the Missouri Department of Natural Resources also sent a joint letter to the Pentagon requesting "top priority funding" from the army in order to hasten the cleanup of FUSRAP sites. The agencies noted that "the potential exposure and movement of contaminated materials" was of "grave concern."[19]

A Long Time Coming

The army did not respond for seven months. When Assistant Secretary of the Army Jo-Ellen Darcy replied in April 2015, she did not commit to additional funding but urged officials to share their concerns with the Army Corps of Engineers' St. Louis District. Darcy noted that "St. Louis sites were given every consideration in the formulation of the Fiscal Year 2016 Budget, along with many other worthwhile programs, projects and activities across the Nation in competition for limited Federal resources."[20] Rene Poche, chief of public affairs for the corps' St. Louis District, later stated that the district's allocation was approximately $34 million for 2016—almost one-third of the $104 million allocated to the corps for the year's entire FUSRAP program. "It should be more than enough to do what we need to do this year," he said.

Beyond funding was the issue of *where* the military was conducting testing and cleanup activities. In 2013, the Coldwater Creek Facts group discovered the Army Corps of Engineers had confined its remediation work to the general area around SLAPS and Latty Avenue. No tests had been conducted to determine whether residential areas north of Interstate 270 were contaminated from past creek flooding. With this new information, Wright went to the trailer at one of the corps' work sites. Holding a map that identified the homes of 750 people who had died or were sick in locations near the creek, Wright said, "Show me there are no radionuclides in the area and I'll shut up." Residents continued the pressure until late 2014, when the corps agreed to begin testing areas north of the highway. By December 2015, engineers had discovered low-level radioactive contamination in twelve sites along Coldwater Creek, including several residential yards, public parks, and commercial sites. For the first time in its fifteen years of FUSRAP work in the region, the Army Corps of Engineers had confirmed radioactive contamination on residential properties.[21]

"This was huge," Wright explained. "It proved what we were saying all along. But they wouldn't have found it if we hadn't managed to convince them to test this area." By 2016 the army had tested more than ten thousand samples north of Interstate 270 along Coldwater Creek and its ten-year floodplain. Working downstream, the corps continued to locate and remove contamination from many years of flooding. Engineers' efforts were limited to areas within the creek's ten-year floodplain—which some residents deemed inadequate, given the frequency and extent of past flooding. Residents also noted that Coldwater Creek had been diverted in some places since 1946, leaving behind "broken trails." Wright explained, "The dirt has been moved around [by the creek] so much that radionuclides are sure to have migrated much further." She added, "We are fearful that there are many areas that are constantly being used by people that are going to go undetected."[22]

Poche noted that the Army Corps of Engineers employs the "follow the radiation method" to determine the next areas to test. For instance, if a sample in a certain area registers positive for radiation, the corps tests other samples in the vicinity. "We are going to go wherever the sampling leads us to," explained Poche. He indicated that sampling extends beyond the ten-year floodplain only if evidence of further contamination is established.[23]

Citizen concerns about "broken trails" are validated by a map of tributaries off the main channel of Coldwater Creek. They would have received sediment moving downstream—for example, during significant flooding in 1957 and thereafter. However, the branches are largely nonexistent on current maps of Coldwater Creek published by many government agencies, media outlets, and private companies. They depict Coldwater Creek as a single blue line winding from the airport to its mouth at the Missouri River.[24]

Like the army, the Centers for Disease Control took a long time (more than a year) to become involved with radioactive contamination at Coldwater Creek. However, in November 2015 the CDC announced that scientists from its Agency for Toxic Substances and Disease Registry (ATSDR) would be teaming up with county health officials to assess health risks from radioactive contamination in and around the creek. The results of these efforts were released in 2018 and finalized in 2019.

Vindication

The ATSDR found that radiological contamination in and around Coldwater Creek "prior to remediation activities" could have increased the risk of some types of cancer in area residents. The agency stated in its report, "Children and adults who regularly played in or around Coldwater Creek or lived in its floodplain for many years in the past (1960s to 1990s) may have been exposed to radiological contaminants." Such exposure (from soil, sediment, or surface water) may have increased the risk of certain cancers (bone, lung, leukemia, skin, or breast). By contrast, contact from 2000 on may have increased the lifetime risk of bone or lung cancer.[25] The report cites evidence of thorium-230 "above FUSRAP remedial goals" in several areas of the creek's floodplain. In addition, the document notes that since the waste entered the creek "decades ago," the "detailed information" on how it moved with sediment into the floodplain does not exist. Nevertheless, the reporting of the historical use of creek sediment or floodplain in other locations "indicates a possibility that contamination spread from the floodplain."[26]

In its conclusions, ATSDR voiced support of ongoing efforts to identify and properly remediate radiological waste around Coldwater Creek. The report recommends that FUSRAP continue to investigate and clean up Coldwater Creek sediments and floodplain soils to meet regulatory goals. To increase knowledge and allay community concerns, the agency favored the future sampling of areas where soil or sediment had reportedly moved from the creek floodplain. (Examples are "fill" relocated during construction and sediment from flooding of the creek's major residential tributaries.) Additionally, ATSDR recommended the sampling of indoor dust in homes where yards have been cleaned up, and of soil that remains in basements from past creek floods.[27]

While ominous, these findings and recommendations vindicate the Coldwater Creek Facts group and other citizens who shared their initial suspicions. Wright stated, "We're really pleased that the federal government has performed the public-health assessment and worked earnestly to gather the facts and try to assist our community. . . . They listened, they researched, and didn't sweep this thing under the rug."[28]

The ATSDR report also recommends that exposed or potentially exposed residents share its findings with their physicians as part of their medical history and consult with the physicians promptly if unusual symptoms develop.[29] The St. Louis County Department of Public Health immediately followed the report's publication with a health advisory to all St. Louis–area physicians and oncologists. The notice gave a summary of results, as well as contact information for support in caring for patients with a history of exposure.[30]

Ray Hartmann, publisher of *St. Louis Magazine*, authored an article with the headline "We Wrote about Poisons in Coldwater Creek 37 Years Ago. Guess What the Feds Just Confirmed?" He expressed outrage by reprinting a portion of his 1982 piece in the *Riverfront Times*: "There is absolutely no excuse for the dangerous byproducts of World War II's military effort to be threatening a major population center 37 years later." Hartmann observed, "Someone should be responsible for seeing to it that radioactive waste be kept safely stored away from such centers and the waterways that serve them. That 'someone' is unmistakably the federal government."[31]

Noting his earlier observation that young children are taught to clean up their own mess, Hartmann added, "Someone should send the Department of Energy to its room."[32] Other publications also covered the work of the Coldwater Creek Facts group, including newspapers (the *St. Louis Post-Dispatch*, *St. Louis Business Journal*, and *Wall Street Journal*) and news organizations (St. Louis Public Radio, NBC News, and CBS News). Coldwater Creek Facts also drew the attention of special publications like the *St. Louis Record* (on legal issues) and *Earth Island Journal* (on environmental issues). Finally, many people learned about the nuclear contamination of the St. Louis area from watching the HBO documentary *Atomic Homefront* or the film *The First Secret City*.[33]

Some area residents were disappointed that the ATSDR didn't connect an increased risk of cancer to individual cancer cases. Instead, the agency's report left people with unanswered questions about the exact cause of their illness. Dr. Adetunji Toriola, an epidemiologist at Washington University's Institute for Public Health, explained that a large-scale study of individual cancer cases "would involve interviewing people that had lived within the creek, following them up, looking at their histories and determining their cancer rates." Such a labor-intensive process would be difficult to implement.[34]

Other residents wondered whether there was a cancer cluster in the neighborhood but discovered that obtaining definite proof would be difficult. While the World Health Organization and EPA state that long-term exposure to low-level nuclear radiation is linked to an increased risk of cancer, establishing a cancer cluster is a complicated procedure. The American Cancer Society states that one in three people will develop the disease in their lifetime. While health departments across the United States receive an average of one thousand reports annually indicating possible existence of cancer clusters, most are not investigated because they fail to meet the very specific definition of that term. The CDC and the National Cancer Institute define a cluster as "a greater-than-expected number of cancer cases that occurs within a group of people in a defined geographic area over a period of time." Of the cases investigated, only 10 percent are believed to occur in higher-than-expected numbers, and even then, detailed inquiries frequently fail to identify specific causes for the illness.[35]

With respect to the Coldwater Creek illnesses, "What we can say is that there is a concern but there is no evidence of a cancer cluster yet," explained Dr. Toriola. Nevertheless, medical experts agreed that the reported number of appendix cancer cases around Coldwater Creek (forty-five in a population of about eight hundred thousand) seemed abnormally high. (This rare disease is usually seen in one in every five hundred thousand Americans.) "But there's so much to piece out that it's difficult to say what's going on there at the moment," noted Toriola. "It's going to take a large and concerted effort to figure that out."[36]

"We Are Not 'Activists'"

Although government agencies eventually played a role in identifying cancer risks in north St. Louis County, there is no doubt that Coldwater Creek Facts was pivotal in pushing them to do so. In trying to educate officials, area agencies, and the public about the environmental crisis at hand, the group used a variety of means to advance its goals, including social media, statistical analysis, press contacts, and multiple meetings with government officials. By finding needed information, members could then apply, analyze, synthesize, and evaluate their data and share it with their community. These were acts of environmental education.

One important way Coldwater Creek Facts communicated with constituents was through its web page. The site displays health maps, along with the seven-part 1989 series on radiological contamination published in the *St. Louis Post-Dispatch*. Coldwater Creek Facts also introduced itself to web page visitors, specifying what the group *was* and *was not*:

> We are not "activists"; we are not being paid for what we do. We are not a nonprofit organization, and we are not receiving funds from politicians, private activism groups, and/or donors. We are purely volunteers: a grassroots effort with no hidden agenda. Everything we do is paid for out of our own pockets, and using our own personal time. We are all parents and most of us work outside of the home. We all started out as North County kids, and most of us worked our way through high school and college. . . . Through a lot of hard work (and student loan debt), many of us went on to pursue advanced degrees in science, nursing, statistics, economics, accounting, and political science.[37]

The group's descriptions of members' (mostly) modest backgrounds and its disavowal of activism echo the expressions of other Americans who have unexpectedly faced an environmental crisis. After the 1971 nuclear reactor accident at the Three Mile Island (TMI) power plant in Pennsylvania, the utility sought to reopen the facility fourteen years later. Believing the nuclear industry had once lulled them into a false sense of security, community members rejected promises from the utility and NRC to heed the accident's lessons. Speaking for the group, the Dauphin County commissioner told the NRC, "We are just plain folk from central Pennsylvania. We work hard and worship. We parent and play. We aren't hysterical. We are quite sane and our judgment is that we would rather live without TMI."[38] Interestingly, some local women who engaged with important scientific and medical issues during the TMI debates initially downplayed their competence. One described herself as a "dippy housewife." Another minimized her grasp of complex material by saying, "If a housewife can understand it, anybody can."[39]

In 1978, women protested at Love Canal, a suburban development in Niagara Falls, New York, where a toxic dump containing one hun-

dred thousand leaking barrels of chemical waste anchored their neighborhood. Like the advocates at Coldwater Creek and Three Mile Island,
they had initial difficulty getting officials to listen, and they employed
a range of tactics. Unlike Coldwater Creek campaigners, the women at
Love Canal were largely homemakers with no college education, public
relations experience, or administrative background. Nevertheless, they
formed the backbone of a formidable grassroots movement. Historian Richard S. Newman contends that their path to activism stemmed
from their domestic roles. He explains, "Most neighborhood women
became activists . . . because they felt that their sphere of influence—
the *home*—was under siege."[40]

In studying women's experience with environmental justice and
activism, researcher Karen Bell finds they are more likely to experience inequity and have less control over environmental decisions than
men. Bell suggests the injustices occur because women have lower in-
comes and are perceived as having less social status than men.[41] However, many female leaders in the Coldwater Creek Facts group had
respected credentials or a well-paying job. While reluctant to view
themselves as activists, these women displayed confidence in their
skills. Many no longer lived in the creek's watershed but saw themselves as defending present and former neighbors, friends, and family
members with whom they maintained contact through social media
and other venues.

Just Moms

Only a few miles from Coldwater Creek, a nonprofit organization
called Just Moms STL was founded in 2013 in Bridgeton, Missouri,
by Karen Nickel, Dawn Chapman, and Beth Strohmeyer. The women
were concerned about health hazards posed by the West Lake Landfill, contaminated in 1973 through the illegal dumping of radioactive
waste from Latty Avenue. The site was placed on the Superfund National Priorities List in 1990.[42] Because its radioactive waste was not
generated by the Manhattan Engineer District or the Atomic Energy
Commission between the 1940s and the 1960s, the West Lake Landfill
is not a FUSRAP property. Its cleanup is handled by the EPA rather
than the Army Corps of Engineers.[43]

West Lake's radiological materials are mixed with landfilled refuse,
debris, soil, and fill, and appear in both surface (the upper six inches
of ground) and subsurface (seven to twelve feet or deeper) areas of the

site.[44] Within recent years, residents have been concerned with what appears to be a subsurface landfill fire burning one thousand feet from the radioactive wastes. Technically, it is a chemical reaction called a "subsurface smoldering event"—and not a fire. This self-sustaining, high-temperature reaction deep inside the waste pile is consuming the refuse and causing the landfill's surface to settle. The reaction releases a nauseating smell that penetrates the air for many miles.[45] Members of Missouri's congressional delegation have introduced legislation (with bipartisan support) to shift West Lake cleanup responsibilities to the Army Corps of Engineers. The proposal, favored by Just Moms STL, remains stalled. In February 2018 EPA proposed a partial excavation of contaminants at the West Lake Landfill to remove a "majority" of radioactive material.[46] Just Moms supports a full excavation of contaminants with storage at an out-of-state licensed nuclear facility.[47]

On one hand, this group's name evokes the "plain folk" characterizations of the Dauphin County commissioner and the Coldwater Creek Facts website. However, the name also suggests female members who are protective of families and interested in justice, as was the case at Love Canal and Three Mile Island. Some Just Moms members also belong to Coldwater Creek Facts, but most participate in only one of the environmental advocacy groups.[48]

Peaks and Valleys

In 2013, the Missouri Coalition for the Environment gave the R. Roger Pryor Citizen Activist Award to two recipients. Despite a reluctance to embrace the title of "activist," the Coldwater Creek Facts group received the first award for their efforts to "research, organize, inform and mobilize action on the issue of radioactive waste contamination and related cancers in the Coldwater Creek area." The second award went to the West Lake Landfill Group (now Just Moms), where Facebook page administrators "leverage social media and community organizing to protect and inform the community about Cold War–era radioactive wastes" at the landfill.[49]

Despite these accolades, cleaning up the environment has sometimes proved contentious. Tempers have flared at public meetings where some citizens have pressed Army Corps of Engineers members on the health effects of radioactive contamination; other residents have worried about their property values or simply expressed frustration. Some have sought buyouts of their homes, while others have sued for damages.[50] Because they face strict legal requirements, plaintiffs generally have not prevailed in these cases. Missouri has a five-year statute of limitations for personal injury. Moreover, beyond showing that defendants caused them to be exposed to radiation that exceeded the federal standard of a maximum permissible dose, plaintiffs must demonstrate that the exposure caused their injuries. These matters are difficult to prove.[51]

One exception to this litigation pattern occurred in June 2018 when the Bridgeton Landfill (a former permitted solid waste landfill at the West Lake location) agreed to settle a $16 million lawsuit brought in 2013 by the Missouri Attorney General's Office. The suit claimed a "subsurface reaction was harming the health of nearby residents." Under the terms of the agreement, the landfill owners will pay $12.5 million into the Bridgeton Landfill Community Project Fund. The money will be used only as "compensation and restitution" to communities within a four-mile radius of the shuttered landfill and to pay for environmental cleanups "explicitly authorized by Missouri law." Both Karen Nickel and Dawn Chapman of Just Moms STL were present in court when the agreement was reached. Nickel called the settlement "bittersweet" and noted, "It's time to move forward." The U.S. Environmental Protection Agency approved a revised remedial-design work plan for the West Lake Landfill on November 15, 2019.[52]

On March 18, 2022, the EPA announced the delay of the West Lake Landfill waste removal project after more nuclear waste was found at the site. Responding to concerned residents and political leaders, the agency announced plans for an April 2022 meeting in Washington "to clarify our path forward to closure and to commit to steady, timely progress toward the cleanup." Dawn Chapman urged the EPA to hold a public forum in Bridgeton, stating, "We deserve to sit there and hear the truth about what we've been allowed to live next to for almost 50 years in this community."[53]

Currently, Coldwater Creek (from Banshee Road at the airport to the Missouri River) is one of the FUSRAP properties and is undergoing remediation by the Army Corps of Engineers. FUSRAP sites are associated with uranium, thorium, and radium and their decay products, according to the corps. Importantly, these locations are contaminated with "low levels of residual radioactivity" because the raw material with high-level radioactivity was shipped off-site at the time of processing. The corps emphasizes that none of the FUSRAP sites pose an immediate health risk to the public or the environment *given cur-*

rent land uses. Nevertheless, the materials will remain radioactive for thousands of years and could pose a risk if land uses change. Each site is remediated for the foreseeable future. In general, the existing contamination is "several inches to several feet below ground level, capped with vegetation, asphalt or concrete and/or is in areas that are restricted from the general public."[54]

As of December 2021, the corps has taken more than twenty-nine thousand samples from the Coldwater Creek floodplain, proceeding nearly ten miles from the airport to Old Halls Ferry Road. Test results will suggest a plan for removing subsurface contamination from the watershed. The expected completion date is 2038.[55]

Mary Shaw, sixty-four, who raised children near the creek, observed, "It's been a very long process. It's just been ridiculous. They should have bought us all out." Christi Oster Evans, 58, grew up near Coldwater Creek and relocated from the area. With a healthy lifestyle and no family history of cancer, Oster wondered whether the creek is connected to her recent lymphoma diagnosis. "I didn't think in a million years that this was going to happen to me," she said. "In the last 45 days, my life is like somebody took the rug out from underneath me and shook it."[56]

In assessing current health risks in the Coldwater Creek watershed and elsewhere, it is important to recognize the extent of radiological contamination in our lives. In fact, cancer cannot be attributed to one cause, but rather to the cumulative effects of toxins in the environment. Biologist Sandra Steingraber explains, "From dry-cleaning fluids to pesticides, harmful substances have trespassed into the landscape and have also woven themselves, in trace amounts, into the fibers of our bodies. This much we know with certainty. . . . We should understand the lifetime effects of these incremental accumulations."[57]

Joseph Masco describes the effect of the U.S. nuclear program in more sobering terms. He notes, "The atomic bomb is now deeply embedded into the earth system itself, leaving radioactive traces of twentieth-century nuclear nationalism in literally every ecosystem and living body on earth."[58]

CHAPTER 6

An Environmental
Justice Watershed

All citizens of the United States shall have the same right, in every
State and Territory, as is enjoyed by white citizens thereof to inherit,
purchase, lease, sell, hold, and convey real and personal property.

—**Civil Rights Act of 1866**

Although the area drained by Coldwater Creek is not an official cancer cluster, the EPA designated it as an "environmental justice watershed" following the 2000 census. That designation—based on the number of low-income and minority residents—qualified the region to receive financial and strategic assistance to address environmental and public health issues.[1] EPA's label signaled a change from the 1950s through the early 1970s, a period when the watershed had a largely White and middle-income population. However, the term "environmental justice watershed" raises the question of whether race was a key factor in locating SLAPS. Since several studies have found instances of environmental racism in situating toxic waste sites, SLAPS's history can inform these discussions.

This chapter will present the history of African Americans in the Coldwater Creek watershed, arguing that they faced a clear and pervasive pattern of racial discrimination. However, the 1946 decision to locate SLAPS adjacent to the municipal airport was a result of factors other than race, as described in chapter 2. That being said, racial discrimination affected where African Americans lived in the St. Louis area, as well as when and how they were exposed to radionuclides from the Manhattan Project. The path to becoming an environmental justice watershed began in the 1970s, when White residents began leaving the area—prompted by racial prejudice, worries about property values, quality-of-life issues, and (after the public became aware of radioactive contamination) environmental and health concerns. These points will be explored below.

The Debate

Over the years, scholars have debated why toxic wastes are located in some types of communities more than others. In 1983 Robert D. Bullard reported that solid waste sites in Houston were not randomly scattered but often found in predominantly Black neighborhoods. He described institutionalized discrimination in the housing market, a lack of zoning, and decisions by public officials over time as major influences that caused Houston's Black neighborhoods to become "dumping grounds" for the area's solid waste.[2]

In 1983 the U.S. General Accounting Office documented racial disparity data from four commercial hazardous waste facilities in the Southeast. (Three of the four surrounding populations were largely African American.) In 1987, a report of the Commission on Racial Justice of the United Church of Christ concluded that communities with active hazardous waste facilities consistently had higher percentages of racial minorities than those without such facilities. (Three out of every five Black and Hispanic Americans lived near at least one uncontrolled abandoned chemical waste site.) The report's authors found it unlikely that these patterns were due to random chance and inferred deliberate racial discrimination on the part of the polluters.[3]

On the other hand, in 1994 the Social and Demographic Research Institute at the University of Massachusetts found no statistical difference between the percentages of minorities living in neighborhoods with commercial hazardous facilities and the percentages of minorities living in neighborhoods without them. That year legal scholar Vicki Been looked at "locally undesirable land uses" (LULUs) in minority communities and contended their presence did not necessarily mean that original siting decisions had been made unfairly. Been argued that the presence of a LULU in a neighborhood often causes property values to fall—which, in turn, changes the local demographics.[4]

Inspired by Vicki Been's work, Thom Lambert and Christopher Boerner conducted a 1997 study of environmental inequities in the St. Louis metropolitan region. According to these authors, determining the cause of environmental inequity is "more complex" than calculating demographic data for populations near toxic sites. To establish whether industrial plants were "sited proportionately," one must know "the demographic and racial conditions of their host communities *at the time the facilities were built.*" Lambert and Boerner argue that it is not unusual for the demographics around industrial plants to change over time; indeed, "facilities originally sited in white areas may eventually become surrounded by minority residents." The authors conclude that those who think "a disproportionate percentage of minorities and poor people around industrial plants indicates discriminatory siting" ignore alternative explanations for observed disparities.[5]

Lambert and Boerner point out that each year 17 to 20 percent of Americans move to a new home, largely out of dissatisfaction with their living conditions. An industrial site or waste in a neighborhood may cause it to be perceived as "less desirable," prompting real estate prices to fall and people to relocate. Over time, the authors suggest, the attractiveness of cheap housing would draw lower-income homeowners and renters who "voluntarily" move into the area.[6]

Lambert and Boerner further contend that, since "industrial siting tends to beget further industrial siting," additional new plant construction could likewise prompt lower real estate values and resident relocation. The presence of pollution could have a similar effect, especially after 1970 when Americans developed a heightened environmental consciousness.[7]

Lambert and Boerner's article drew a response from historian Andrew Hurley. In his "Floods, Rats, and Toxic Waste: Allocating Environmental Hazards since World War II," Hurley states, "[While the authors] may be correct in asserting that pollution was partly responsible for white flight, their contention that African American in-migration was voluntary ignores the racist dynamics of the housing market in postwar St. Louis."[8]

By 2018, Vivien Hamilton, Brinda Sarathy, and Janet Farrell Brodie would argue that "white flight, racial segregation, political marginalization, and institutional racism" resulted in low-income communities of color being "disproportionately impacted by toxic spaces" in the United States.[9]

Historic Black Communities

A pattern of racial discrimination in the St. Louis area began well before the post–World War II period. Black people were among the first nonindigenous residents of the Coldwater Creek watershed. French settlers brought enslaved persons of African heritage to the area before the Louisiana Purchase. Thereafter, Black slaves traveled from the

Upper South to Coldwater Creek with their White American owners. During Missouri's territorial and early statehood periods, Black slaves worked on plantations such as Hazelwood and Taille de Noyer (Walnut Grove, in Florissant) while also laboring on farms in the watershed's northern reaches.[10] Interestingly, Hazelwood encompassed land that later became Latty Avenue, an area contaminated with radionuclides from SLAPS. Established by Richard Graham in the early nineteenth century, the plantation's name is credited to Graham's friend Henry Clay—former Speaker of the House, U.S. senator from Kentucky, and secretary of state. Clay jokingly suggested naming the estate after the proliferation of hazel bushes growing there. Graham liked the idea, and the name remained.[11]

In 1828, there were 2,439 people living in Old St. Ferdinand Township, which encompassed the Coldwater Creek watershed. Of that total, 496 were either enslaved or free persons of color. By 1860, the township's Black population was 863, and by 1870 it was 952. Census records show that for both years, the number of Black people in Old St. Ferdinand Township surpassed that in other St. Louis County townships, as well as in every ward in the City of St. Louis.[12]

Over time, several Black communities formed in the vicinity of Coldwater Creek. In Ferguson, Thomas January gave his former slaves land near the intersection of Eddy Avenue and Florissant Road. These families established the Mount Olive Baptist Church as the neighborhood's center. The congregation drew additional members from a Black settlement one mile southeast of Taille de Noyer, near the intersection of Powell and Elizabeth Avenues.[13]

Another Black community was Robertson, begun in the early nineteenth century by French-speaking settlers and their slaves. After the Civil War, many emancipated persons remained in the area. African Americans also lived in the northeastern part of Bridgeton, once called Marais des Liards ("cottonwood swamp") by the colonial French. Black people worshipped at Bridgeton Baptist Church (established 1853) and an African Methodist church (established 1874) and built a school. In 1868, freed Black persons from Bridgeton and Black Jack started a one-room church and school on Old Halls Ferry Road. In 1886 they collected money from tenant farmers and sharecroppers to establish an African American cemetery called the New Coldwater Burying Ground that remained active until 1949.[14]

The community of Breckenridge Hills had an early African American neighborhood that originated with Washington Reed, a sergeant in the U.S. Colored Artillery during the Civil War. After his honorable discharge Reed married, worked on rented farms, operated a dairy, and taught Black children. By 1889 he had saved enough money to purchase forty-four acres of land. His heirs subdivided the property in 1948 amid the post–World War II housing boom.[15]

Eventually these Black communities disappeared. By 1980, only one family remained on Ferguson's Eddy Avenue. Since 1950, an elementary school has sat on the Powell Avenue site where Black families once lived. Interstate highway construction and airport expansion took Bridgeton's African American neighborhood; Robertson also fell to airport expansion. In the late 1920s, Black Jack's church and school on Old Halls Ferry Road burned to the ground.[16]

Today, one historic Black community still exists in the Coldwater Creek watershed, albeit as a shadow of its former self. The community is Kinloch, developed in the 1890s and incorporated in 1948.[17]

Kinloch

Kinloch Park began as a commuter suburb for White residents, with a small section reserved for African Americans employed as servants. The racial line became blurred when "Mrs. B" and her husband (who were Black) reportedly bought property in the White section of Kinloch Park through their friendship with the White owner. (A restrictive covenant prohibited the sale of property to Black people within the designated White section.) Upon learning of the purchase, White residents moved out of Kinloch Park and were not replaced by new White buyers. The Olive Street Terrace Realty Company decided to develop a suburb for African Americans by selling lots for five dollars down and five dollars per month. When the firm was unable to use Black persons' notes of purchase as collateral for bank loans, it changed course. The company found White purchasers, who resold the lots to Black buyers at inflated amounts—usually double the initial price. Within a few years, some thirty Black families moved into the five or six blocks in the southeast portion of the Kinloch Park, which became known as South Kinloch Park.[18]

In 1902 Kinloch established a public school system from parts of two adjacent districts. Consistent with state law, Black and White students were enrolled in separate schools. In time, the district's African American population increased, surpassing that of Whites. When a Black minister won election to the three-person school board, his advocacy for a Black high school concerned White property owners. (The only Black high school in St. Louis County was thirteen miles away and difficult to access.) Fearing that voters would elect a second Black board member, White residents incorporated as the City of Berkeley, and then acted in accordance with a state law that allowed a municipality to establish its own school district. Since Berkeley's boundaries were drawn to include White people and exclude Black residents, the downsized Kinloch district had a reduced tax base that made adequate funding of its schools very difficult.[19]

Historian Colin Gordon reports that Kinloch was isolated and shunned by its neighbors—even though one-third of St. Louis County's African American population resided there before 1960. He reports that Kinloch enjoyed few of the advantages of its White suburban peer communities and that the zoning and planning history of neighboring Berkeley and Ferguson "were largely animated by the desire to quarantine Kinloch and its residents." Gordon adds, "Most Berkeley and Ferguson streets dead-ended before they reached Kinloch, and until 1968 Ferguson barricaded the through streets."[20]

Census records show that Kinloch's population declined after reaching a peak of 6,501 in 1960.[21] In 1976, the community's racially isolated school district and the neighboring Berkeley School District were merged with the larger Ferguson-Florissant system under federal court order.[22] Then, beginning in 1982, the "Lambert–St. Louis International Airport" began buying (under an FAA noise abatement program) 1,040 of the 1,404 parcels that had once made up Kinloch.[23] Between 1990 and 2000, Kinloch lost more than 75 percent of its population, with many displaced residents moving to nearby north St. Louis County communities.[24] Today only 300 people live in Kinloch. Due to the airport noise abatement buyout, the City of St. Louis owns 406 of the 1,080 parcels of land there; St. Louis County owns Kinloch Park and "a handful of vacant properties that were put on the tax rolls," according to the *St. Louis Post-Dispatch*. The City of Kinloch owns just 211 parcels. Kinloch has become an illegal dumping ground where junk piles and broken mattresses litter the streets. Residents are calling for the City of St. Louis, now the largest landowner, to lead the cleanup.[25]

Locating SLAPS

In 1946, Robertson, Bridgeton's Black community, and Kinloch were within five miles of the 21.7 acres that became SLAPS. (The approximate distances were 1.5 miles, 2 miles, and 4.5 miles, respectively.) However, the White communities of Berkeley, Ferguson, and Bridgeton were larger and located similar distances from the site. The owners of the condemned property were White, as was a majority of the surrounding rural population. Importantly, when the United States filed a condemnation suit to acquire SLAPS, officials did not envision a typical solid waste disposal site. They wanted a place to "store" uranium processing residues and wastes for another nation, under the strict secrecy of the Manhattan Project. Expediency, security, and a political connection to the City of St. Louis were important to the MED. For their part, St. Louis officials likely appreciated that residents living around the site weren't city voters. Beyond these considerations, citizens of north St. Louis County were of little concern in locating SLAPS.

City and County

Beginning in the postwar years, new developments were shaping the Coldwater Creek watershed. The 1960 census showed that outmigration had increased St. Louis County's population to 703,532—just short of St. Louis City's 750,026. Suburban growth in the watershed was dizzying. Florissant had a 1,000 percent increase in population to 38,166. Berkeley tripled its total to 18,676. Bridgeton grew from 202 to 7,820. St. Ann increased from 4,577 to 12,155, and Ferguson and Overland each doubled to nearly 23,000 people.[26] By 1970, the county's population would surpass the city's (961,353 to 622,236).[27]

The net result, according to the 1960 census, was a 12.5 percent population loss for St. Louis, with some spaces offset by new arrivals from poor rural locales. During this period the city's Black population increased 28.8 percent; however, its neighborhoods were generally not diverse. The north side of St. Louis was solidly Black; the central corridor and near south side were mixed; and the south and southwest areas were predominantly White. James Neal Primm contends the pattern was intentional—as seen, for example, in city planners' placement of public housing on the north side.[28] His assessment recalls earlier instances of St. Louis officials locating "nuisances" among vulnerable populations through selective law enforcement and through the pioneering use of zoning laws.

Other practices upheld the hegemony of race and class in St. Louis-area neighborhoods. Well into the late 1940s, real estate boards considered Black occupancy a nuisance. After 1950, they dropped an explicit racial reference for language that cautioned agents against introducing "a character of property or use" that harmed neighborhood property values. According to Gordon, real estate boards continued to base home values on class or "racial homogeneity of the neighborhood" as much as on physical structure. In 1962, the American Institute of Real Estate Appraisers defined "neighborhood" as an area exhibiting "a fairly high degree of homogeneity as to housing, tenancy, income, and population characteristics."[29]

Gordon notes that although there were many restrictive covenants designed to accomplish the same goal, the White Real Estate Exchange "also regulated and constrained its members even where no covenants existed." Segregation of the real estate profession continued well into the 1960s, with Black realtors belonging not to the exchange but to the separate Real Estate Brokers Association.[30] In the late 1950s, St. Louis newspapers still listed rental and resale properties available to African Americans under the separate heading "for colored."[31] Gordon notes that in St. Louis County, "beyond Kinloch and the transitional inner-ring suburbs, patterns of African American settlement were watched closely."[32]

Race and class concerns motivated the move to the suburbs. However, while city dwellers once looked down on county residents during the Great Divorce, county dwellers now disparaged city folks. Race concerns were paramount, as underscored by a 1947 pamphlet extolling the virtues of suburban life in St. Louis County. The booklet rates the neighborhoods according to average rents, population density, and "presence of negroes" while stating, "People Who Can, Move Away."[33]

Jones v. Mayer

Eventually, the desire of African Americans to move beyond closely prescribed areas resulted in a landmark decision of the U.S. Supreme Court. Clarence Harmon (who later became St. Louis's second Black mayor) recalled trying to buy a home in Paddock Woods in the late 1960s. It was a development of the Alfred Mayer Company, St. Louis County's largest homebuilder. "Well, sir, we are not selling homes to Negroes up here," Harmon remembered hearing. "There's a case coming up before the Supreme Court, and they will determine whether we sell to any Negroes."[34]

The plaintiff was Joseph Lee Jones, who in 1965 attempted to buy a Paddock Woods home. When the Alfred Mayer Company turned Jones away because he was Black, he filed a complaint in district court and lost. The court held that the developer's refusal amounted to private discrimination and not a "state action." That position was supported in an appeal to the circuit court. However, in 1968 the U.S. Supreme Court struck down the prior rulings on the last day of Earl Warren's term as chief justice. The decision in *Jones v. Mayer* cites the Thirteenth Amendment guarantee (supported by the Civil Rights Act of 1866) that "all citizens of the United States shall have the same right, in every State and Territory, as is enjoyed by White citizens thereof to inherit, purchase, lease, sell, hold, and convey real and personal property." The Supreme Court's action reinforced Title VII of the 1968 Civil Rights Act (passed earlier that year), which broadly prohibits discriminatory practices by realtors, developers, and lenders. Following the Supreme Court's decision, Clarence Harmon moved into Paddock Woods.[35]

Jones v. Mayer helped to open homes to Black buyers. By 1990 there were 26,984 African Americans living in municipalities of the Coldwater Creek watershed. The numbers of Black residents per municipality were as follows: Berkeley, 8,164; Black Jack, 2,697; Breckenridge Hills, 1,001; Bridgeton, 651; Ferguson, 5,589; Florissant, 2,078; Hazelwood, 1,625; Kinloch, 2,684; Overland, 1,147; St. Ann, 943; and St. John, 405.[36]

Ticking Time Bombs

African Americans who moved to the Coldwater Creek watershed were subject to a phenomenon described by historian Andrew Hurley. He notes that several St. Louis–area chemical companies and heavy industries abandoned toxic waste sites where they once operated facilities. Although these sites were largely located in the city or "heavily industrialized suburbs," few of them were in Black neighborhoods when the hazardous substances were introduced. However, as African Americans replaced White working-class residents, the newcomers were unaware of the "environmental timebombs" ticking in their backyards.[37] In a similar manner, African Americans who moved into the Coldwater Creek watershed, likely with high hopes for the future, were unaware of the existence of radiological contamination when they purchased their homes. In that way they resembled my White family, who optimistically moved to the area in 1957.

Black Experiences / White Flight

The movement of African Americans into the Coldwater Creek watershed suggests they did not uniformly experience radioactive contamination from SLAPS. Some residents of historic Black communities received early, dangerous, and direct exposure to the contaminants (for example, members of the Robertson Fire Department, as described in chapter 2). Others moved into the watershed after *Jones v. Mayer* and subsequently learned that "environmental timebombs" were already in their backyards, as Hurley reported. Still others arrived in the area after 2000 and eventually discovered, through the ASTDR report, that their chance of contracting certain types of cancers differed from that of people who lived near Coldwater Creek in the late twentieth century.

Residents of the watershed—including those living there before and after 2000—often expressed their concern for property values. Sometimes they communicated the fears in racial terms (reflecting the views of realtors who presented negative images of Black people). However, after the public learned about radiological pollution in the area, growing numbers of residents worried about its effect on home resale values. This is evident in press accounts, the *St. Louis Site Remediation Task Force Report*, and questions residents posed to the Army Corps of Engineers.[38]

The reason for these concerns is clear. For Americans, homeownership has long been the primary means of building wealth. Predictable house payments create stability. A stable home (absent frequent relocations) fosters academic achievement in children. By contrast, losing one's home equity can spell disaster and prevent the transfer of wealth to the next generation.[39] In addition to property values, environmental and quality-of-life issues surfaced as the Coldwater Creek watershed became more densely populated. This phenomenon is seen, for example, in the wistful description of suburbanization's effect on Black Jack, Missouri, and the government reports on growing industry in the area.[40]

Prompted by these factors, White residents began to exit the Coldwater Creek watershed from the 1970s onward. Departures due to race were likely accelerated by the *Jones* decision and the federal court mandate that merged the Ferguson-Florissant and Berkeley school districts with the all-Black Kinloch system.[41] Florissant's population reached its peak in the mid-1970s; Hazelwood's and Berkeley's peaked in the 1990s. While all three communities have continued to lose people, largely White St. Charles County (to the immediate west) had a 90 percent population increase between 1990 and 2020.[42]

In 2014, the Missouri Department of Health and Senior Services discovered the importance of acknowledging White flight in assessing the effects of radiological contamination at Coldwater Creek. One year prior, the department had conducted a study that compared the cancer cases in four area zip codes against the expected number between 1996 and 2004. According to the analysis, it is unlikely that radiation exposure increased the incidence of cancer. When Coldwater Creek Facts members criticized the study for failing to consider substantial population turnover in the watershed, the department redid its calculations and arrived at very different conclusions. This time the division found higher-than-normal rates of some cancers, supporting the position of the Coldwater Creek Facts advocacy group.[43]

Current Municipalities

The 2020 population figures demonstrate how much Coldwater Creek's watershed has changed since 1946, when the population was rural—or even since 1960, when White middle-income residents were in the majority. Table 1 presents the fifteen current municipalities in the watershed according to a variety of measures. Seven of the municipalities are situated upstream from SLAPS; the other eight are located downstream from the site where most of the contamination spread.

Among the downstream municipalities, all had substantial African American populations in 2020, ranging from 26 percent in Bridgeton to 82.27 percent in Berkeley. Excepting Florissant and Calverton Park, all downstream communities had poverty rates that exceeded the U.S. average of 10.5 percent. Home values in all 8 communities were below $167,700, the average home price in Missouri. Only Bridgeton and Hazelwood exceeded the median U.S. household income of $68,703. In 5 communities (Berkeley, Bridgeton, Ferguson, Hazelwood, and Kinloch), the percentage of residents who own their homes fell below the U.S. average of 64.3 percent.[44] The numbers appear to support what James Neal Primm said shortly before his death in 2009: "The wealthier population of St. Louis has always been running from poverty."[45]

TABLE 1
Municipal Comparisons in the Coldwater Creek Watershed, 2020

Municipality	Year of Incorporation	Total Population	% African American Population	Median Home Price	% Homeownership	Average Household Income	% in Poverty
Breck. Hills	1950	4,553	34.96	$73,500	31.5	$54,498	27.05
Bridgeton*	1843	11,514	26.00	$161,700	63.5	$82,197	12.71
Calv. Park*	1940	1,267	52.57	$84,500	68.4	$65,789	8.70
Edmundson	1948	830	27.37	$64,700	39.6	$49,757	20.79
Ferguson*	1894	20,500	69.93	$81,600	54.0	$52,886	21.59
Florissant*	1786	51,750	38.25	$96,400	65.3	$63,615	9.93
Hazelwood*	1949	24,974	35.85	$116,800	59.5	$73,030	11.74
Kinloch*	1948	289	79.71	NA	23.0	$26,305	50.36
Overland	1939	15,461	22.92	$87,000	60.2	$55,296	15.70
St. Ann	1948	12,573	30.32	$87,000	54.5	$52,173	15.24
St. John	1945	6,517	24.0	NA	NA	$46,313	12.30
Sycmore Hls	1941	655	27.34	$98,400	83.7	$62,697	16.64
Wdsn. Terr.	1954	4,028	34.17	$69,800	51.9	$54,604	12.31

sources: World Population Review (worldpopulationreview.com/us-cities), Surburban Stats (SuburbanStats.org)
* signifies downstream location (below SLAPS)

Risky Business

Following the release of information by public health officials and the Army Corps of Engineers, citizens living in the Coldwater Creek watershed have expressed misgivings about their health while worrying about home values. Hazelwood resident (and cancer patient) Mary Oscko discussed her home's resale prospects after the Army Corps of Engineers found radioactive contamination in the neighborhood. She asked, "Who would buy it?" and "Who would I in good conscience sell it to?"[46] The current modest home values around Coldwater Creek seem to validate Oscko's concerns.

Some area residents wonder about the impact of radioactivity on future generations. They say that while the Army Corps of Engineers seeks to make their environment safe *for current land uses*, the contaminants will be around for a long time. People wonder whether present restrictions will be forgotten many years hence, thereby endangering future populations.

Beneath all the issues is an important question concerning environmental justice in the Coldwater Creek watershed: Who is ultimately responsible, and are they being held accountable?[47] Although posed by one resident at a public meeting with the Army Corps of Engineers, this question is on the minds of many. In the meantime, disclosures of radiological contamination in the Coldwater Creek watershed have not slowed population losses. The revelations have provided one more reason to move, for those who could afford it.

CHAPTER 7

A Part of the Whole

We have the oldest radioactive waste of the atomic age. And
there is no place on the planet to put this where it won't impact
our air, our water, and our lives. There is no solution.

—Kay Drey, St. Louis environmental activist, 2013

Several members of Coldwater Creek Facts reported having personal experience with chronic low-level ionizing radiation. While in his thirties, political science professor Scott McClurg developed a brain tumor after eating vegetables from his Florissant garden. He became the lead plaintiff in a suit against Mallinckrodt and other businesses.[1] Alicia Helbling, who lived near Coldwater Creek throughout childhood, developed a rare salivary gland tumor. (Her mother died of brain cancer at age thirty-nine.) "The age [of sick people] really stuck out to me," Helbling recalled, adding, "I discovered that my rare tumor is commonly seen in Hiroshima bomb victim survivors."[2] Kim Thone Visintine, a Coldwater Creek Facts founding member, lost her six-year-old son Zach to brain cancer. Registered nurse Shari Riley noticed a striking number of cases of appendix cancer in north St. Louis County (hers among them). She took the lead in organizing the cases, contacting each person individually and offering support. Riley succumbed to the disease in 2014.[3]

An Emerging Narrative

To comprehend the magnitude of these losses, it is important to recognize that events at the Mallinckrodt Chemical Works and the Coldwater Creek watershed *were only part* of an enormous U.S. nuclear weapons program operating under the Manhattan Project and its successors. Between 1945 and 1990, the United States produced more than seventy thousand nuclear weapons at a cost of roughly $409 billion.[4] The Department of Energy lists twenty-nine facilities throughout the country that participated in nuclear weapons research, testing, and production during those years.[5]

Some public records of the program are, if not actively misleading, remarkably opaque. The DOE lists the innocuous-sounding Destrehan Street Plant—the uranium refining factory at Mallinckrodt Chemical Works that sent radioactive wastes to SLAPS. Also included are the Paducah Gaseous Diffusion Plant, which produced enriched uranium; the Rocky Flats Environmental Technology Site, which made plutonium-239 "pits" to trigger nuclear fission; and the Hanford Reservation, which manufactured plutonium for the Nagasaki bomb.[6]

Each of these facilities is the subject of an article or book written by someone who grew up near the site. According to Brinda Sarathy, Vivien Hamilton, and Janet Farrell Brodie, they represent an emerging narrative that shows how the global nuclear complex has affected individuals. Stories within the narrative explore how scientists and politicians made decisions on where to dump nuclear waste. These accounts offer personal testimony on what it was like to grow up near a secret weapons facility, or work in a plutonium plant, or survive a nuclear disaster. The stories also give voice to the "struggle for recognition and reparation . . . for multiple communities impacted by the nuclear industry globally."[7] As such, they follow the groundbreaking work of Terry Tempest Williams, who in 1991 wrote *Refuge: An Unnatural History of Family and Place* after discovering her mother, some relatives, and community members had cancer they linked to nuclear testing.[8]

In 2000, short story writer Bobbie Ann Mason published an article in the *New Yorker* titled "Fallout: Paducah's Secret Nuclear Disaster."[9] Twelve years later, Kristen Iversen—director of the MFA program in creative writing at the University of Memphis—wrote the nonfiction *Full Body Burden: Growing Up in the Nuclear Shadow of Rocky Flats*.[10] In 2020, attorney Trisha T. Pritikin wrote *The Hanford Plaintiffs: Voices from the Fight for Atomic Justice*.[11] The similarities and differences of those narratives—with one another, and with developments at the Coldwater Creek watershed—offer insight on how the nuclear complex impacted local communities.

Paducah, Kentucky

Mason's article describes Paducah's support of the Gaseous Diffusion Plant even after the reporting of radiological contamination there. The small city's most famous resident, U.S. vice president Alben W. Barkley, used his political clout to locate the complex on the site of an old munitions factory. Although the new facility processed and enriched uranium, it was locally known as "the bomb plant." To Paducah residents, it was a good neighbor that brought well-paying jobs to the community. They knew uranium was dangerous but didn't expect to be harmed by it. They also believed that keeping a secret was something they owed their employer (30).

When the *Washington Post* (and then the *Louisville Courier Journal*) disclosed radiological contamination at the bomb plant in 1999, the Paducah newspaper remained silent. Outside news organizations reported on radioactive waste dumps, safety violations, bureaucratic lies, increasing occurrence of cancer, and environmental pollution, but Paducah residents were unimpressed. ("If Tom Brokaw came, then it would be real," one citizen observed.) After visiting the site, Mason corroborated some of the allegations by describing "blue cylinders of depleted uranium, rows and rows of them" that would remain there "until someone [figured] out an economical way to recover the last valuable traces." She also reported an atmosphere of denial, as when the management said one employee's sickness was caused by eating too much country ham. (His body was exhumed several years after his death, and the autopsy showed uranium levels hundreds of times above normal.) When a deer in the nearby wildlife area was found to contain plutonium, the Kentucky state health commissioner falsely assured citizens they would have to "eat the whole deer" to experience any significant effects (31–34).

Although the bomb plant was not supposed to handle plutonium, it arrived anyway and remained there. Citizens' worry increased when they learned an accidental uncontrolled nuclear reaction was theoretically possible at the plant. Eventually a $10 billion class action lawsuit was filed against former contractors Union Carbide and Lockheed Martin. Energy Secretary Bill Richardson visited the plant; he apologized for the plutonium and "promised the moon," according to Mason. However, residents generally remained passive during the ordeal. "It was more than high-paying jobs," she explained. This was a community where the social contract was "'going along by getting along'" (36).

Jefferson County, Colorado

Kristen Iversen grew up in Bridledale, an Arvada, Colorado, neighborhood that represented "the golden dream of suburban life and all
its postwar promises." It was near the Rocky Flats Nuclear Weapons
Plant, which mass-produced the fissionable plutonium pits at the core
of nuclear bombs. The AEC owned Rocky Flats, which it described as
"a virtual waste land" chosen for "operational values." The site was on
a high, windy plateau not far from the growing cities of Boulder and
Denver. The federal government acquired the land by eminent domain and rushed the facility's construction. Dow Chemical operated
the plant and was indemnified against any accident or mishap. Later
Rockwell and EG&G, formally known as Edgerton, Germeshausen,
and Grier, Inc., managed the operation (*Full Body Burden*, 5–7, 9).

Unfortunately, the AEC failed to consider important environmental factors in locating the new plant. The site criteria specifically stated
that wind passing over the facility should not blow toward any major
population center. However, engineers based their analysis on wind
patterns at Stapleton Airport on the opposite side of Denver, where
winds come from the south. Rocky Flats, by contrast, is subject to "chinook" winds traveling down the eastern slope of the Rockies from the
west and northwest—"straight toward Arvada, Westminster, Broom-
field, and Denver." One engineer noticed the error and warned against
the location because Denver was downwind a few miles away. He was
ignored (Iversen, 6–7).

Iversen explains that the passage of the Atomic Energy Act of 1946
created an "impenetrable wall of secrecy around the U.S. nuclear establishment." All government decisions and activities related to the production of nuclear weapons were hidden. Information about nuclear
bombs, waste, environmental contamination, and health risks was
strictly classified. In their public announcements, AEC officials assured
people that Rocky Flats would not emit dangerous wastes. Nor would
atom bombs or weapons be built there—just some unspecified component parts. Iverson notes that plutonium pits created at Rocky Flats
formed the explosive fissionable core essential to every nuclear weapon
in the U.S. arsenal. Each individual pit could function as an atomic
bomb "of the same type as the Trinity and Nagasaki bombs" (4–6, 18).

Other problems at Rocky Flats stemmed from the enormous quantities of waste created by mass production. Beginning in 1954, thirty-
to fifty-gallon drums began to stretch "nearly as far as the eye [could]

see," exposed to the elements. Each barrel held waste oil and solvents contaminated with plutonium and uranium. They couldn't be shipped or stored, and no on-site building could hold all of them. Over time, the contaminants leaked into the soil and groundwater that fed into two nearby creeks, a lake, and a reservoir. Dow knew about the leakage but kept it secret, according to Iversen. Eventually the drums were sent to a waste site in Idaho, while others were buried at Rocky Flats, contaminating the groundwater. Even after the drums' removal, wind continued to scatter plutonium for miles (Iversen, 48, 66, 77).

In 1957 a fire in the plant consumed some $20 million worth of plutonium. AEC noted in a classified report that 41 employees had endured exposure to substantial doses of radiation in the fire. Both AEC and Dow publicly said no contaminants had been released off-site, and the public was never at risk. Nevertheless, subsequent tests revealed that plutonium had spread over the area. One official claimed that a person would have to "eat the dirt"—and large amounts of it—for plutonium to be dangerous (Iversen, 30–43, 66–67).

In fact, plutonium is dangerous if ingested or inhaled. Its particles can lodge in lung tissue for decades, creating ionizing radiation. One nuclear chemist who worked at Boulder became suspicious and began to collect soil samples around the plant to examine for evidence of plutonium-239 and strontium-90. His results caused the AEC to send its own scientists, who found plutonium from Rocky Flats on and off the site, covering an area of over thirty square miles. Nevertheless, the AEC insisted the local population was safe. The nuclear chemist became concerned about the two hundred thousand to three hundred thousand people who lived immediately downwind from Rocky Flats, especially in Denver's suburbs. He estimated that a potential nuclear disaster at the plant could devastate Denver and possibly all of Colorado. However, there was no emergency response plan to protect the public during a major disaster at Rocky Flats, which remained "the biggest secret in town" (Iversen, 59–67).

Despite these issues, Rocky Flats was a boon to the Denver area. Hundreds of millions of federal dollars were pumped into the local economy through salaries and commercial contracts. The population grew as developers plowed contaminated soil beneath the ground's surface for new buildings, creating breathable dust. However, as more information surfaced on nuclear contamination of the area, home prices fell and people worried about their property values. In 1974 a citizen task force convened to seek answers to growing questions about

Rocky Flats. Its report concluded that there were serious safety issues at the plant, as well as the potential for a nuclear accident there. In a press conference, one task force member stated that Dow was "neither responsible nor responsive" to the public or Rocky Flats; moreover, the company's secrecy was supported by the AEC. Dow Chemical left Rocky Flats after two decades of oversight as Rockwell stepped in to carry on "business as usual" (Iversen, 102, 105, 123–124, 204).

As small groups of activists began to appear at Rocky Flats' gate, Dr. Carl Johnson, Jefferson County health director, used funding from the National Institutes of Health to study locations downwind from Rocky Flats. He and his colleagues found an average of forty-four times more plutonium in nearby soil than had been reported to the State Department of Health. Concentrations in the air and drinking water were elevated. Moreover, there were higher-than-average rates of cancer (primarily leukemia and lung cancer) in areas downwind from the plant (130). Amid growing citizen outrage, a date was set for the first big national protest at Rocky Flats: April 29, 1978. Continuing demonstrations generated national publicity, with Daniel Ellsberg among those arrested (Iversen, 144–57).

Then in 1980, EPA admitted to the press for the first time that cancer deaths from Rocky Flats contamination might occur among Denver residents. The conclusion was based on reports of the Department of Energy (AEC's successor) and the fact that air-monitoring stations at Rocky Flats consistently showed higher levels of plutonium-239 than elsewhere in the Western Hemisphere (Iversen, 168).

This was unwelcome news for the county commissioners, county board of health, DOE, and Rockwell. They were focused on business and had no interest in hearing about contamination, according to Iversen. The county commissioners (who monitored business growth and new home development) appointed the county board of health, which, in turn, appointed the health commissioner. Rockwell received bonuses from the DOE based on production, and not on health improvements. Since the Atomic Energy Act of 1954 exempted nuclear weapons plants from environmental laws, Rockwell and DOE wanted to regulate the facilities themselves (Iversen, 169).

To keep Health Commissioner Johnson from releasing alarming reports to the press, the county board of health added the president of the homebuilders association to its group. Two weeks later the board voted to demand Johnson's resignation. He was given the choice of being fired and losing all his benefits or resigning immediately. The seven-year veteran resigned and filed suit (Iversen, 169–70).

Meanwhile, Rocky Flats' problems were mounting. For decades the plant had been shipping its waste to a "temporary" storage facility at the Idaho National Engineering Laboratory. Iverson writes that, in the high desert location, "millions of cubic feet of plutonium-contaminated waste" sat in rusting barrels, drums, and boxes, "waiting for the federal government to find a permanent dump site." The Idaho venue was not safe, being situated above the Snake River Plain aquifer, in an earthquake zone, and on a floodplain. The site held 3.5 million cubic feet of plutonium that is not expected to stabilize for 240,000 years (Iversen, 195). Although U.S. officials had promised that the Idaho dumping ground would be used only until 1970, when the waste would be moved to a new location, Iverson states there was no new site. The Water Isolation Pilot Plant near Carlsbad, New Mexico, once viewed as a contender, had its own "technical, structural and management deficiencies (Iversen, 195–96).

On June 26, 1989, more than 90 FBI agents and EPA investigators raided Rocky Flats, seizing thousands of documents and hundreds of samples of waste. The evidence indicates that for more than thirty years, "spills, leaks, and waste disposal practices" had contaminated "dozens of sites" around the plant. The final FBI and EPA allegations include "concealment of environmental contamination, false certification of federal environmental reports, improper storage and disposal of hazardous and radioactive waste, and illegal discharge of pollutants into creeks that flow to drinking water supplies" (Iversen, 225–26).

Rockwell officials had no comment on specific federal allegations at Rocky Flats. Plutonium operations ceased after the raid, and on November 21, 1990, President George H. W. Bush declared the end of the Cold War (Iversen, 239–40). Five years later, due to its proximity to Denver, the DOE named Rocky Flats the most dangerous facility in the U.S. Cold War complex of nuclear weapons production sites.[12]

Kristen Iversen, however, contends that Rocky Flats was not unique. She asks, "Is it fair to single out Rocky Flats?" and adds, "Plenty of other DOE facilities, including Hanford in Washington State, Oak Ridge in Tennessee, the Savannah River site, and the Fernald plant in Ohio have severe problems with radioactive and toxic waste and storage. Some are worse than Rocky Flats."[13]

This brings the discussion to the secret city that became famous for producing plutonium for Fat Man—the bomb dropped on Nagasaki that ended World War II.

Hanford, Washington

The Hanford Engineer Works, originally called the Hanford Site, produced the plutonium for both the Trinity test and the Nagasaki bomb. Michele Stenehjem Gerber notes that despite Hanford's "enormous engineering, chemical, and atomic puzzles," its greatest challenges were environmental. The Columbia River Basin, the site of the Hanford project, had "loose, porous soil, . . . flat, open topography, . . . dry winds, and . . . big, swift rivers" that presented significant problems for site operators.[14]

A search for the site began in December 1942, when Lieutenant Colonel Frank Matthias of the Army Corps of Engineers traveled with DuPont executives to locate the world's first industrial-scale plutonium plant. Operating under the Manhattan Project, the corps would supervise the facility, with DuPont serving as the primary contractor.[15] Before undertaking the work, DuPont insisted on full indemnification from the U.S. government.[16]

Matthias settled on the Hanford area of Washington State because it had several critical features: plenty of clean water from the Columbia River, a reliable source of electricity from the Grand Coulee Dam, a high percentage of government-owned land, and "a certain scent of failure." He rejected the lush wheat fields and prosperous-looking farms closer to the dam in the belief that farmers would not easily (or cheaply) relinquish them. Matthias found the Hanford area "far more promising" because of the sparse population, meager crops, and "shabby look" of its ranches. According to Kate Brown, Matthias came as an "un-founding father" to the "luckless communities" he selected for removal. Federal appraisers offered reimbursements that generally fell short of the worth of the year's requisitioned crops—not to mention the value of the real estate. Property owners had no recourse but to "pack up and file a petition in court, hoping to get a better price."[17]

Lindsey A. Freeman contends that the acquisition of land at Hanford was a repeat of the procedure followed at Oak Ridge, Tennessee: "Kick out the people, clear the land, and bulldoze it to the point where newcomers might even suspect they had landed in a territory that had never been settled before." While General Groves first thought that work on uranium and plutonium could occur in one spot, Enrico Fermi's successful experiment at the University of Chicago on December 2, 1942, changed Groves's mind. Fermi demonstrated the need for a huge area to locate reactors to produce the needed amount of plutonium. Not wanting to push Oak Ridge beyond its limits, and feeling the area was too populated to risk dangerous plutonium production, the general "looked westward for a plutonium-focused site." The choice was Hanford.[18]

The first construction at Hanford was a camp established in 1943 to house workers for a plutonium plant. It had all the charm of a medium-security prison, according to Brown, and "everybody was spied on."[19] When the camp of single male transient workers took on the "boozing, brawling" ambience of a frontier town, DuPont executives decided to build a new operators' village "dedicated to workers safely rooted in nuclear families."[20] By mid-1945, Hanford Camp had been bulldozed.[21]

After the Army Corps of Engineers flattened the original ranch town of Richland (while also relocating Native American tribes and depopulating the town of White Bluffs), military officers and DuPont executives worked to repopulate it.[22] The two groups decided on a community largely inhabited by White middle-class families, with working-class residents displaced to the cultural margins. Bankrolled by federal subsidies, the upscale, exclusive bedroom community became a model for a middle-class suburb—a novel concept in 1944 that would be replicated many times over in the postwar era.[23]

By the spring of 1945, the corps officers and corporate managers at Hanford had accomplished a great deal. Within a period of two years, they had built a series of factories and the world's first industrial reactors for plutonium production. They had also demolished three towns and built in their place two new cities and a labor camp from the ground up.[24] Gerber notes that Hanford was characterized by several features, chief among them being "speed and boldness, innovation, secrecy, and the strict yet oblique handling of radiological hazards in the workplace." The nearby public "was not so well protected nor informed as were Hanford workers," in Gerber's view.[25]

Meanwhile, Manhattan Project researchers were becoming uneasy as they learned more about radioactive isotopes. Scientists noticed the way they quickly lodged in the body and lingered there, affecting biological systems. The scientists' fear, according to Brown, was that the radioactive isotopes "would no longer be an external feature of human existence, but . . . a lasting detour (or cul de sac) on the path of human evolution."[26]

Believing the problem could be averted by minimizing human contact with these isotopes, researchers looked for the paths by which plutonium and other fission products might enter the body. Upon finding they migrated outdoors—to grasslands, into rivers, and into air currents—the researchers pondered Hanford's location and drew inferences described by Brown:[27]

> The idea in locating the Hanford plant in the wide-open, sparsely populated Columbia Basin was to use the local territory as a vast sink into which engineers could dispose of hundreds of thousands and eventually billions of gallons of radioactive and toxic waste. With a vast reach of territory, the scientists figured, radioactive isotopes would scatter into the air, soil, and water to the point where they would be so diluted as to be harmless everywhere to everybody. The strong winds would carry away radioactive gases from high smokestacks. The swift, high-volume Columbia River would speed off liquid waste to the Pacific Ocean. The earth in the miles-wide buffer zone around the plant and the sandy sediment under the plant would easily absorb radioactive waste and make it vanish.[28]

These assumptions were not borne out in fact. As engineers moved to dispose of radioactive and toxic waste in the environment, some residents noticed changes that caused concern.

Attorney Trisha T. Pritikin reports that, in the mid-1950s, families began arriving at the local hospital with children who were critically ill with leukemia. Tests revealed that area youngsters exhibited high levels of a radioactive isotope of iodine (I-131) from the local milk. In the winter of 1961, farmers living near the Hanford plant noticed that new lambs were born dead, stunted, or deformed (*Hanford Plaintiffs*, 5, 25, 127). Then, in the mid-1970s, Juanita and Leon Andrewjeski observed that many young men in their farming community suffered from heart attacks. When Leon was also found to have heart disease, Juanita tracked the occurrences, listing the names, ages, and addresses of afflicted neighbors and plotting the locations on a map. The number of "X"'s (for heart attacks) and "O"'s (for cancer) grew to "alarming proportions." A family friend contacted Karen Dorn Steele, a reporter for the *Spokane Spokesman-Review*, and she toured the area that was known as the "death mile" (Pritikin, 200–201).

In 1986 the federal government allowed the mass declassification of documents related to the Hanford Site. Four years later, a government report (*Initial Hanford Radiation Dose Estimates*) confirmed what many people had suspected. Beginning with the start of operations in 1944 and for more than four decades thereafter, the Hanford nuclear facility secretly released ionizing radiation throughout vast areas of the Pacific Northwest—affecting eastern Washington, Idaho, western Montana, northern Oregon, and downriver communities along the Columbia River. With public knowledge of the report, almost five thousand people who lived downwind and downriver from the plant turned to the courts for relief. They filed personal injury claims in mass toxic tort litigation against the facility's former contractors, claiming their cancers and other serious illnesses were a result of exposure decades earlier when they were in utero—or during their infancy or childhood (Pritikin, xi, xiii, 238).

Pritikin was one of the Hanford "downwinders" who claimed to have been exposed to ionizing radiation prior to birth. (Her mother drank milk every day during pregnancy, "faithfully following the latest government-issued nutritional guidelines for a healthy baby.") As a result, Pritikin suffered from thyroid disease and hypothyroidism, and eventually needed a total thyroidectomy. She states that declassified documents reveal the AEC and its successor, the DOE, concealed decades of off-site radiation releases from the Hanford facility. Officials failed to monitor exposures to Hanford's downwind and downriver communities and repeatedly assured residents they were in no danger when radioactive fallout blanketed the community. Pritikin's oral history is one of twenty-four statements included in her book *The Hanford Plaintiffs: Voices from the Fight for Atomic Justice* (xiii, 23, 253).

The Hanford plaintiffs were especially upset to learn of the Green Run experiment on December 2–3, 1949, that deliberately released 5,500 curies of iodine-131 and other fission products into the atmosphere. There was no public health warning prior to this event. Iodine-131, a product of plutonium production, is absorbed by the thyroid gland and can cause cancer or other malfunctions decades later.[29]

The Hanford downwinders' case lasted twenty-five years, with many plaintiffs dying before a final settlement was reached. The United States paid an undisclosed amount for the downwinder agreement, as well as over $80 million in legal fees to law firms defending the nuclear contractors. This action was necessitated by a World War II agreement indemnifying the contractors for the risky and uncertain business of making plutonium.[30]

Pritikin concludes her book with a question: "What about the others?" She observes that Hanford is not the only Manhattan Project or Cold War nuclear weapons facility that released airborne radiation offsite. Probing further, she asks, "What about communities exposed to radioactive waste from the Mallinckrodt Chemical Works, which processed uranium for the U.S. nuclear program and stored waste aboveground north of St. Louis, Missouri?"[31] In fact, Mallinckrodt's refining of uranium and storage of wastes in the Coldwater Creek watershed can be compared to the three facilities discussed in this chapter.

Comparisons

The nuclear sites at Paducah, Rocky Flats, Hanford, and Mallinckrodt / Coldwater Creek share many commonalities. All four operated under a cloak of secrecy. Moreover, officials at each facility ignored important environmental principles in choosing its location. In Paducah, Alben W. Barkley selected the polluted site of a former ordnance factory in order to stimulate the economy of his hometown. Rocky Flats' selection stemmed from a misunderstanding of wind patterns in the Denver area. Hanford scientists wrongly assumed their plant's location would cause the river, sand, and air to absorb radioactive wastes and make them vanish. The SLAPS at Coldwater Creek was also a poor choice for storing radioactive materials. One-third of the site was located on a floodplain. In addition, its downward slope drained into the tributary, which carried radionuclides 13.7 miles to the Missouri River.

Officials at each location also failed to adequately prepare for the enormous quantities of waste their industrial plants would produce. At Rocky Flats and Coldwater Creek, rusting drums of radioactive waste sat aboveground as far as the eye could see. Paducah had thirty-seven cylinders of depleted uranium and "a giant mound" of crushed fifty-five-gallon drums—all within the New Madrid earthquake zone.[32] Hanford emitted its wastes into the atmosphere, jeopardizing the health of downwinders throughout the Pacific Northwest. At both Rocky Flats and Coldwater Creek, "temporary" waste storage sites remained far beyond their intended stay for want of a good alternative.

All four installations provided a discernable economic benefit to some constituency. Mallinckrodt uranium workers received wages that exceeded the going rate, as did workers at Paducah and Rocky

Flats. Since in all cases nuclear energy production helped the local economy, the advantages discouraged action from present and former employees who discovered radiological contamination in their midst. For example, Bobbie Ann Mason's sister worked at the Paducah plant for several years. When confronted by reports of radiological contamination, the sister replied, "I guess I was exposed. . . . But don't worry. If you got it you got it, and there is nothing that can be done—but maybe it can for the next generation." The sister reminded Mason of how good the plant had been to employees, with its high salaries and excellent benefits. She added that the environment was "so secret" that you felt you were doing something important for the country.[33]

Disclosures of contamination and safety violations often came from people who did not work at the facilities—journalists, scientists, or area residents who noticed an unusual amount of illness or death around them. Sometimes they conducted their own investigations, as did the Boulder scientist who collected soil samples, or Juanita Andrewjeski and the Coldwater Creek advocates who compiled their own health maps.

People who uncovered safety violations or health hazards were often met with denials from those in charge. Sometimes officials confirmed a small amount of contaminant while greatly understating the threat it posed to human beings. Examples include the Kentucky official who tried to explain away a deer contaminated with plutonium, or Missouri and Colorado officials who told residents they would have to "eat the dirt" to be affected by radionuclides. Occasionally officials tried to silence the messenger, as when members of the Jefferson County board of health and the homebuilders association colluded to fire a health commissioner for exposing hazards at Rocky Flats.

Over time, concerned residents became more involved by forming citizens' groups, participating in protests, or filing lawsuits. Sometimes federal agencies intervened, pointing fingers at one another. Examples include the EPA action against the Department of Energy in the West Lake Landfill, and the 1989 FBI-EPA raid on Rocky Flats, operated by Rockwell under Department of Energy ownership.

Despite the four sites' similarities, there were also distinctions between them. The facilities were in different regions of the country, in varying communities. Paducah was a small city in the Upper South; Rocky Flats was near the Denver suburbs; Hanford was surrounded by an agricultural land in the Pacific Northwest. Mallinckrodt Chemical Works was in a large industrial city in Missouri. The Coldwater

Creek watershed evolved over time from a rural locale, to a suburban setting, to an urban area.

There was also a difference in the public's awareness of each site. Despite the secrecy, Paducah residents recognized the bomb plant. Jefferson County inhabitants could identify Rocky Flats, even if some thought it manufactured cleaning supplies. Denizens of Richland, Washington, and the surrounding countryside had no trouble pointing to the Hanford Engineer Works looming on the horizon. By contrast, the St. Louis Airport Storage Site was not a recognizable presence to most people in the Coldwater Creek watershed. They would have known Mallinckrodt as a St. Louis company (one among many), located eleven miles away. They also would have been familiar with the St. Louis airport at Lambert Field but were likely unaware of the 21.7-acre storage site on an adjacent property. Even after the February 1989 series in the *St. Louis Post-Dispatch* about nuclear contamination in the St. Louis area, few residents of Florissant, Hazelwood, or Black Jack would have worried about their exposure to SLAPS. For that reason, McCluer North High School alumni did not initially suspect that Coldwater Creek was making their friends and loved ones sick. These advocates had to do the research, design and administer a survey, and analyze the results before settling on the creek that flowed through their neighborhood as the cause of these ailments. The threat hid in plain sight for decades.

Yet the fact that the Coldwater Creek watershed shared many commonalities with Paducah, Rocky Flats, and Hanford means it is not an outlier. Rather, it is part of a larger whole created by the Manhattan Project and its successors.

Then and Now

There were five people in my family of origin who moved to Florissant in 1957, as described in the introduction. Among the group, only Jimmy and I would have full adult life spans. My father, mother, and brother John all died of lymphoma at early ages—fifty-two, fifty-six, and sixty-three, respectively.

Dad, Mom, and John never knew that radionuclides in Coldwater Creek posed a threat to their health. During their lifetimes, reporting mechanisms were not sufficiently developed to keep track of cancer incidence, especially in a mobile population. Federal legislation did not make cancer a reportable disease until 1973, or mandate that all cancer cases be disclosed to state agencies until 1992. Until 1985 there was no National Cancer Database. In the mid- to late 1980s, 15 percent of hospitals were not in compliance with new cancer-reporting laws.[34] Certainly my relatives' illnesses and deaths could not have shed light on a possible cancer cluster in our former neighborhood.[35] However, their experiences, presented below, suggest that radiological contamination in Coldwater Creek has affected area residents for a long time.

My father worked near the St. Louis Airport Storage Site in 1947–48 and again from 1957 until his death. In his final fifteen years, he resided in the Coldwater Creek watershed. He received a cancer diagnosis in St. Louis County in 1971 and died there the following year. My mother lived in the Coldwater Creek watershed from 1957 to 1977. She received a cancer diagnosis in the City of St. Louis in 1977 and died in another state the same year. John resided in the Coldwater Creek watershed from 1957 to 1971. He moved out of state, where he was diagnosed with cancer in 2007 and died in 2009. My husband, Jim (who lived in the Coldwater Creek watershed from 1948 to 1973), was diagnosed with cancer in the City of St. Louis in 1982. Thirty-six years later, he received a different cancer diagnosis in St. Louis County, where he continues to undergo treatment.

Looking back, I often recall the words of my uncle Herman, a Wichita physician who in 1977 visited our Florissant home during my mother's final days. "I don't believe St. Louie is a very healthy place to live," he told me. "Everyone on the street has some damn tumor."

When I recall my youth in Florissant, fond memories are now tinged with ambivalence and perhaps some irony. Coldwater Creek was part of the landscape of my childhood. My brothers searched for crawdads there, and I played softball at a park that abutted the flowing stream. However, now I understand these "good" events occurred in the midst of a "bad thing." The radionuclides were there all along; we just didn't know it. Aside from the actual loss of my parents and brother, the most painful part of my association with Coldwater Creek is revising the warm recollections of childhood to accommodate a grim reality. As a former U.S. history teacher, it is also unsettling to acknowledge the likelihood that my loved ones' deaths were part of the cost of World War II. Hazelwood resident Mary Oscko put it succinctly: "We are the casualties of the Second World War. This mess is because they wanted to win and went and built that bomb."[36]

Evaluation of Community Exposures: Summary

Public Health Assessment for

Evaluation of Community Exposures Related to Coldwater Creek

**ST. LOUIS AIRPORT/HAZELWOOD INTERIM STORAGE SITE (HISS)/
FUTURA COATINGS NPL SITE**

NORTH ST. LOUIS COUNTY, MISSOURI

EPA FACILITY ID: MOD980633176

APRIL 30, 2019

U.S. DEPARTMENT OF HEALTH AND HUMAN SERVICES
PUBLIC HEALTH SERVICE
Agency for Toxic Substances and Disease Registry

THE ATSDR PUBLIC HEALTH ASSESSMENT: A NOTE OF EXPLANATION

This Public Health Assessment was prepared by ATSDR pursuant to the Comprehensive Environmental Response, Compensation, and Liability Act (CERCLA or Superfund) section 104 (i)(6) (42 U.S.C. 9604 (i)(6)), and in accordance with our implementing regulations (42 C.F.R. Part 90). In preparing this document, ATSDR has collected relevant health data, environmental data, and community health concerns from the Environmental Protection Agency (EPA), state and local health and environmental agencies, the community, and potentially responsible parties, where appropriate.
In addition, this document has previously been provided to EPA and the affected states in an initial release, as required by CERCLA section 104 (i)(6)(H) for their information and review. The revised document was released for a 60 day public comment period. Subsequent to the public comment period, ATSDR addressed all public comments and revised or appended the document as appropriate. The public health assessment has now been reissued. This concludes the public health assessment process for this site, unless additional information is obtained by ATSDR which, in the agency's opinion, indicates a need to revise or append the conclusions previously issued.

Agency for Toxic Substances & Disease Registry...................Robert R. Redfield, MD, Director, CDC, and Administrator, ATSDR

National Center for Environmental Health/Agency for Toxic Substances & Disease Registry ...Patrick Breysse, PhD, CIH, Director

Agency for Toxic Substances & Disease Registry..Christopher M. Reh, PhD, Associate Director

Division of Community Health Investigations...Susan Moore, MS, Acting Director

Central Branch ....Jack Hanley, MPH, Acting Chief

Eastern Branch... CAPT Alan Parham, MPH, Acting Chief

Western Branch ....Rhonda Kaetzel, PhD, Acting Chief

Science Support Branch...CAPT Peter Kowalski, MPH, CIH, CSP, Acting Chief

Use of trade names is for identification only and does not constitute endorsement by the Public Health Service or the U.S. Department of Health and Human Services.

Additional copies of this report are available from:

Agency for Toxic Substances and Disease Registry
Attn: Records Center
1600 Clifton Road, N.E., MS F-09
Atlanta, Georgia 30333

You May Contact ATSDR Toll Free at
1-800-CDC-INFO
or
Visit our Home Page at: http://www.atsdr.cdc.gov

St. Louis Airport/   Final Release
Hazelwoold Interim Storage Site (HISS)/
Futura Coatings NPL Site

PUBLIC HEALTH ASSESSMENT

Evaluation of Community Exposures related to Coldwater Creek

ST. LOUIS AIRPORT/HAZELWOOD INTERIM STORAGE SITE (HISS)/
FUTURA COATINGS NPL SITE

NORTH ST. LOUIS COUNTY, MISSOURI

EPA FACILITY ID: MOD980633176

Prepared by the
U.S. Department of Health and Human Services
Agency for Toxic Substances and Disease Registry
Division of Community Heath Investigations
Atlanta, Georgia 30333

Summary

Introduction

The Agency for Toxic Substances and Disease Registry (ATSDR) evaluates community exposures and makes recommendations to prevent harmful exposures to hazardous substances in the environment. This report evaluates potential exposures to people who played or lived near Coldwater Creek in North St. Louis County, Missouri. Historical radiological waste storage sites near the St. Louis Airport released contamination into Coldwater Creek. The Army Corps of Engineers' Formerly Utilized Sites Remedial Action Program (FUSRAP) has been characterizing and cleaning up areas related to these sites since 1998.

Community members asked ATSDR to do this evaluation. They are particularly interested in exposures that occurred in the past, before storage site cleanup began.

This report uses available environmental data and information from the community to evaluate whether people playing or living near Coldwater Creek have or had harmful exposures to radiological or chemical contaminants from the creek. This report also addresses other exposure concerns which could not be fully assessed and makes recommendations for further work.

A draft of this report was provided for public comment from June through August 2018. Changes made in response to public comments are summarized in the report, and detailed responses to comments are provided in an Appendix. Although details of our evaluation changed, ATSDR's overall conclusions remain the same

Conclusions of ATSDR's Evaluation

To evaluate possible effects from exposures, ATSDR estimated the exposure and resulting risks for children and adults who directly touched, swallowed, or breathed in sediment and water from Coldwater Creek and soil in its floodplain for many hours a day for many years. We assumed they were always exposed to concentrations of contaminants present in the most highly contaminated areas. Based on different specific assumptions for past (1960s to 1990s) and recent (2000s and on) exposures, detailed in this report, we reached the following four conclusions

Conclusion 1

Radiological contamination in and around Coldwater Creek, prior to remediation activities, could have increased the risk of some types of cancer in people who played or lived there.

Basis for Conclusion

- Children and adults who regularly played in or around Coldwater Creek or lived in its floodplain for many years in the past (1960s to 1990s) may have been exposed to radiological contaminants. ATSDR estimated that this exposure could have increased the risk of developing lung cancer, bone cancer, or leukemia.
- More recent exposures (2000s and on) only slightly increased the risk of developing lung cancer from daily residential exposure.
- Estimation of risk, especially for past exposures, involved many uncertainties. The estimated increased risks would not likely result in detectable increased cancer rates in the community as a whole.

Next Steps

- ATSDR recommends that potentially exposed residents or former residents share their potential exposure related to Coldwater Creek with their physicians as part of their medical history and consult their physicians promptly if new or unusual symptoms develop. Upon request, ATSDR can facilitate a consultation between residents' personal physicians and medical specialists in environmental health.
- ATSDR recommends that the state consider updating analyses on cancer incidence, cancer mortality, and birth defects, as feasible.
- ATSDR will provide technical support, upon request, to update cancer incidence or mortality studies in the area and identify public health actions needed.

Conclusion 2

ATSDR does not recommend additional general disease screening for past or present residents around Coldwater Creek.

Basis for Conclusion

- The predicted increases in the number of cancer cases from exposures are small, and no method exists to link a particular cancer with this exposure.
- Not all current or former residents would have experienced exposures as high as assumed by ATSDR in this evaluation.
- Screening people who have no symptoms has risks, including false negative results, false positive results, risks from treating cancers that might never have caused a

problem during a person's lifetime, and additional radiation exposure from diagnostic testing. A personal physician will use a patient's individual history, symptoms, age, and gender to determine appropriate screening and diagnostic testing.

Next Steps

- ATSDR recommends that potentially exposed residents or former residents share their potential exposure related to Coldwater Creek with their physicians as part of their medical history and consult their physicians promptly if new or unusual symptoms develop. Upon request, ATSDR can facilitate a consultation between residents' personal physicians and medical specialists in environmental health.

Conclusion 3

ATSDR supports ongoing efforts to identify and properly remediate radiological waste around Coldwater Creek.

Basis for Conclusion

- Thorium-230 (Th-230) has been found above FUSRAP remedial goals in several areas of the Coldwater Creek floodplain. Reducing Th-230 levels in accessible areas will reduce harmful exposures.
- Waste entered the creek decades ago, and detailed information about how it moved with sediment and into floodplain soil does not exist. Reports of historical use of Coldwater Creek sediment and floodplain soil in other locations indicates a possibility that contamination spread from the floodplain. Identifying and remediating contaminated areas outside the floodplain will reduce potentially harmful exposures

Next Steps

- ATSDR recommends that the FUSRAP program continue investigating and cleaning up Coldwater Creek sediments and floodplain soils to meet regulatory goals. To increase knowledge about contaminant distribution and allay community concerns, we recommend future sampling include
    - areas reported to have received soil or sediment moved from the Coldwater Creek floodplain (such as fill used in construction)
    - areas with possible soil or sediment deposited by flooding of major residential tributaries to Coldwater Creek
    - indoor dust in homes where yards have been cleaned up or require cleanup
    - sediment or soil remaining in basements that were directly flooded by Coldwater Creek in the past
- ATSDR recommends signs to inform residents and visitors of potential exposure risks in areas around Coldwater Creek not yet investigated or cleaned up.

- ATSDR will review new data from Coldwater Creek investigations, upon request, and update conclusions if necessary.

Conclusion 4

Other exposure pathways of concern to the community could have contributed to risk. ATSDR is unable to quantify that risk.

Basis for Conclusion

- No sampling data exist that would allow ATSDR to estimate exposures from other pathways, including inhaling dust blown from historical radiological waste storage piles and past consumption of local dairy or agricultural products.

Next Steps

- ATSDR recommends that public health agencies continue to evaluate, to the extent possible, community concerns about exposure and educate the community about radiological exposures and health.
- ATSDR will remain available to provide, upon request, further technical assistance to the public, partner agencies, or other stakeholders.

NOTE

These conclusions may change following availability of new environmental sampling data.

v

APPENDIX B
"Health Advisory"

Steven V. Stenger
County Executive

Faisal Khan, MBBS, MPH
Director

FAX

DATE: June 25, 2018

FROM: Saint Louis County Department of Public Health

TO: Saint Louis Area Primary Care Physicians and Oncologists

TITLE: Health Advisory: Report of Coldwater Creek Community Exposures
             Released

PAGES TO FOLLOW: 3

The U.S. Department of Health and Human Services Agency for Toxic Substances and Disease Registry has released a public health assessment indicating that historical radiological contamination in and around Coldwater Creek in St. Louis County could have increased the risk of some types of cancer in people who have played or lived in and around the creek. A summary of the results of this report for healthcare providers is included, as well as contact information if you need support in caring for patients with a history of exposure.

6121 North Hanley Road • Berkeley, MO 63134 • PH 314/615-0600 • FAX 314/615-6435
RelayMO 711 or 800-735-2966 • web http://www.stlouisco.com/HealthandWellness
An Equal Opportunity Employer - Services Provided on a Non-Discriminatory Basis

Evaluation of Coldwater Creek Community Exposures
North St. Louis County, Missouri – Information for Health Care Providers

ATSDR developed this flier to give health care providers information to help address patients' concerns related to radiological exposure to contaminants from the Coldwater Creek site.

St. Louis was a center for refining uranium ore to support the Manhattan project. Byproduct waste material was stored along the banks of Coldwater Creek (St. Louis Airport, North St. Louis County) for drying the material prior to being shipped for additional processing. It was stored in open piles, and precipitation carried some of the radiologically contaminated waste material into the creek. Flooding carried contaminated sediments to some residential yards. Since 1998, the U.S. Army Corps of Engineers has been characterizing and cleaning up areas related to the site. The Agency for Toxic Substances and Disease Registry (ATSDR) evaluated community exposures to radiological contaminants in sediment, water, or soil while playing or living near Coldwater Creek.

THE BOTTOM LINE – Coldwater Creek Community Exposures

- Thorium-230, and to a lesser extent, radium-226 and uranium-238 are the contaminants of concern.
- Exposures from thorium and other radiological contaminants to residents who played or lived along the creek in the past (1960s to 1990s) may have increased the lifetime risk of certain cancers (bone, lung, leukemia, skin, or breast).
- Exposures to residents who lived along the creek more recently (2000s and on) may have increased the lifetime risk of bone or lung cancer.
- The effective whole-body dose is not expected to result in any harmful health effects and doses to individual organs would not result in any non-cancer health effects.
- ATSDR does not recommend any added general cancer screening for people who played or lived near Coldwater Creek. Patient medical and exposure history and presentation, best practices, and recommendations from the U.S. Preventive Task Force should be used to determine medical management.

Health Effects of Thorium, Radium, and Uranium

Th-230, Ra-226, and U-238 are naturally occurring radioisotopes that emit alpha particles as they decay. The energy of the alpha radiation emitted is similar. All have very long half-lives and will not decay appreciably during a person's lifetime. For Th-230 and Ra-226, radiological effects are expected to predominate (that is, no health effects from their chemical interactions with the body are known to occur before effects from the radiation are observed). Uranium, on the other hand, may cause chemical damage to the kidney microtubules before any radiation effects would be seen. However, uranium doses at Coldwater Creek were many magnitudes lower than concentrations likely to result in health effects.

Thorium, radium, and uranium taken up into the bloodstream are known to build up on bone surface and may be incorporated into the bone matrix. Inhaled thorium, radium, and uranium can also be retained in the lungs. Historical exposures at Coldwater Creek could result in bone surface doses of up to 63,000 millirem (mrem) – 250 times lower than bone surface doses associated with bone cancers. Past exposure could result in red marrow doses up to 3,200 mrem – 50 times lower than red marrow doses associated with leukemia. Past exposure could result in lung doses up to 2,900 mrem – 6,800 times lower than doses associated with lung cancers. Recent exposures were estimated to be lower than past exposures.

Agency for Toxic Substances and Disease Registry
Division of Community Health Investigations

ATSDR

Approach to Patient Management

We recommended in our report that potentially exposed residents or former residents share their potential exposure related to Coldwater Creek with their physicians as part of their medical history and consult their physicians promptly if new or unusual symptoms develop. Upon request, ATSDR can facilitate a consultation between residents' personal physicians and medical specialists in environmental health.

Exposure History

Take an exposure history of patients to learn how long and for how often they may have been exposed. Direct contact with or close proximity to contaminated media near the creek is necessary for exposure. ATSDR's evaluation was based on a worst case very frequent exposure of extended duration.

Radiological Testing

Radiological testing is NOT indicated for the doses potentially received from this site. While radioactive materials can be measured indirectly by analyzing blood, feces, saliva, urine, or the whole body for different types of ionizing radiation, the quantity of the slowly released radioisotopes from the bone would likely be very small and possibly undetectable over instrument background levels.

Cancer Screening

- Screen patients according to standard clinical protocols and U.S. Preventive Task Force recommendations as suggested by the patient's presentation, age, and gender.
- Many procedures that could detect the cancers of interest are associated with risk (such as additional radiation from imaging) that may outweigh the potential benefit.

Advise/Counsel Patients

- As indicated, reassure patients that not all current or former residents would have experienced exposures as high as conservatively assumed by ATSDR in this evaluation.
- Counsel patients that doses received were much lower than doses associated with disease, fertility problems, birth defects, and cancer in studies of radiologically-exposed populations.
- Advise patients that radiological testing or non-symptomatic disease screening beyond standard clinical protocols is not indicated given the exposure doses estimated at this site.
- Inform patients that radiation-induced cancers are indistinguishable from cancers caused by other factors and that it is difficult to attribute any cancer to Coldwater Creek due to the time that has passed and the uncertainty in past exposure estimates.

Where to Learn More

ATSDR Case Studies in Environmental Medicine: www.atsdr.cdc.gov/csem/csem.html
- Taking an Exposure History
- Radon
- Uranium

ATSDR Toxic Substances Portal: www.atsdr.cdc.gov/substances/index.asp
ATSDR Coldwater Creek Site: www.atsdr.cdc.gov/sites/coldwater_creek

For questions about ATSDR activities at this site, contact the site team at ColdwaterCreek@cdc.gov.

These summary tables provide ATSDR's estimated lifetime cancer risks from past (1960s – 1990s) or recent (2000s and on) exposure to contaminants in Coldwater Creek sediment, water, or floodplain soil.

Please see ATSDR's Public Health Assessment at www.atsdr.cdc.gov/sites/coldwater_creek for details.

PAST EXPOSURES (1960s-1990s)

	Number of Lifetime Additional* Cases in 10,000 People		U.S. Lifetime Risk of Specific Cancer in 10,000 People
	Recreational	Residential	
Bone	10	30	10
Lung	3	10	650
Leukemia	1	4	150
Skin	less than 1	2	200
Breast	less than 1	1	1,200 (in women)

RECENT EXPOSURES (2000s and on)

	Number of Lifetime Additional* Cases in 10,000 People		U.S. Lifetime Risk of Specific Cancer in 10,000 People
	Recreational	Residential	
Bone	less than 1	6	10
Lung	less than 1	1	650

Tables show highest estimated increased lifetime risks of cancer at organ site for recreational or residential exposures, assuming exposure from birth to 33 years. Only cancer types with a greater than 1 in 10,000 risk are shown. Approximate U.S. lifetime risk of specific cancers based on SEER data and shown for comparison only.

*U.S. lifetime background risk of all cancers is about 3,800 cases for every 10,000 people.

"FUSRAP Questions":
St. Louis District

In 2016, the U.S. Army Corps of Engineers, St. Louis District, posted a document on its website that contained citizen questions on the Coldwater Creek cleanup, along with answers from the corps. Presented verbatim on the following pages, "FUSRAP Questions" reveals residents' concerns about radionuclides in their neighborhood; it also shows the corps' understanding of, and plans for, the cleanup. Irregularities in numbering, spelling, or grammar are reflections of the original document, which was removed from the website in 2018 when the information was no longer current.

FUSRAP questions

PROCEDURES

1. *Is there an evacuation plan for living too close to the soil?*
   The FUSRAP has no requirement for evacuation while performing remediation, nor have any locations been found to have levels high enough to require evacuation prior to remediation.

2. *Risk Assessment Fact Sheet states that exposure to the creek has a low cancer risk associated with it, according to monitored data collected since 2000. Yet, as you perform soil samples, areas with higher level of radiation continue to be found. How can you reconcile Fact Sheet with what we are seeing now?*
   The levels of radioactive contamination requiring remediation are just above our clean up levels. In its current configuration (below the surface), the contamination does not pose a risk to human health or the environment.

3. *How many mile-radius is effected by the contamination?*
   A mile radius has not been established for the North County

Sites. The USACE continues to sample Coldwater Creek (CWC) and adjacent properties. CWC from Banshee Road to the Missouri River is within the FUSRAP boundaries and is approximately 14 miles in length.

4. *Why haven't you been to my yard?*

   The USACE is sampling CWC and adjacent properties within the 10 year flood plain. If contamination is found, the sampling will continue until the boundaries of the contamination are determined. We are still sampling properties adjacent to CWC to the St. Denis Bridge. We will start sampling from St. Denis Bridge to Old Halls Ferry Road late in 2016. If your property is adjacent to CWC and within the 10 Year Floodplain we will send you a right of entry to sample your property.

5. *Where does the contaminated soil go? How is it disposed of? Are we just moving one problem site to another?*

   The contaminated soil is hauled in covered trucks to the North County loadout area at the St. Louis Airport Site (SLAPS). From there, it is put into covered rail cars and shipped to an out of state licensed disposal facility in Idaho. The facility in Idaho is specifically licensed to receive low-level radioactive waste.

6. *What specific contaminates are you removing?*

   The primary radiological contaminants associated with CWC are radium 226, thorium 230, and uranium 238. Please see Table 2–10 of North County Record of Decision for a more detailed list.

7. *Once you find contamination, what do you do for the homeowners?*

   Once found, we notify the homeowners of the contamination on their property and explain to them what the next steps will be to remove the contamination. The USACE will document the findings in a Pre-Design Investigation Summary Report which will be sent to each property owner. The property owner will be able to review and comment on the document. A Remedial Design will also be written and submitted to the property owner for review and comment. The USACE will then remediate the affected property. The USACE will work with the property owner to remove contaminants from their property to below levels required for remediation. Finally, a Post Remedial Action Report will be written. The property owner will again get the chance to review and comment on the document before being sent a final copy of the document.

8. *You tested my yard. Where are my results?*

Once we have completed the sampling to St. Denis Bridge, we will prepare a document called the "Pre Design Investigation Summary Report" that will contain all the sampling data. After the first of the year, we will be sending out letters to those property owners who have no contamination on their property. If contamination is found on your property, we will contact you in person.

9. *You tested my yard. Please re-test.*

We typically will not re-test a yard unless the first test shows levels of contamination above our remediation guidelines. If we have found levels above guideline limits you will be notified and a plan developed for additional sampling in your yard. If you have a particular concern or new information relevant to your property, please contact our office 314-260-3905.

10. *Is there a list of the sites tested and sample-by-sample results?*

Once we have completed the sampling from Frost Avenue to St. Denis Bridge, we will prepare a document called the "Pre Design Investigation Report" that will contain all the sampling data. This document will be available to the public. After the first of the year, we will be sending out letters to those property owners who have no contamination on their property. If you do have contamination, we will contact you in person to inform you of the presence of contamination on your property.

11. *Is there more information on the website?*

Yes; Newsletters, Fact Sheets, Environmental Monitoring Reports, Pre Design Investigation reports, Remedial Investigation Reports, Feasibility Study reports, Records Of Decision, etc. can be found on the website.

12. *How far apart is the sampling? How is that determined? What affects more/less sampling?*

The USACE uses a nationally accepted assessment protocol called the *Multi Agency Radiation Survey and Site Investigation Manual* (MARSSIM). MARSSIM provides detailed guidance for planning, implementing and evaluating environmental and facility radiological surveys conducted to demonstrate compliance with a dose-or risk-based regulation. The USACE also studies the areas of concern and takes additional samples that are located specifically to evaluate areas with a higher contamination potential such as low-lying areas adjacent to the creek and areas of high sediment deposition.

The Pre Design Investigation Work Plan for CWC from Frost Avenue to the St. Denis Bridge can be found on the St. Louis FUSRAP website.

13. *How do you choose sample locations? Why is Coldwater Creek still accessible?*

See question #12. Possible contaminated areas in CWC display no risk in its current configuration (sub surface).

The Pre Design Investigation Work Plan for CWC from Frost Avenue to St. Denis Bridge can be found on the St. Louis FUSRAP website.

There is no immediate risk to a recreational user from exposure to the NC Record of Decision contaminants of concern (See ROD risk table). There are other risk factors and warning signs posted along the CWC (MDNR and MSD).

14. *How do you test in a home that had storm water / creek water back up through the sewer?*

If a home were to be tested, direct measurements of the potentially impacted surfaces or sediment sampling might be performed.

15. *How do you clean the trucks? Not all your trucks are using the tarps.*

The trucks used to carry contaminated soil to the load out facility are covered. The trucks are visually inspected and scanned before leaving an excavation area. If contamination is found, the trucks are brushed off or hosed down with water and rescanned to make sure no contamination is left.

16. *Shouldn't sites that involve parks (Duchesne, St. Cin) be closed to the pubic?*

The areas where contaminated soils are found and remediated are closed off to the public for the duration of the remediation activities. The decision to close parks is entirely up to the city where the park is located.

St. Cin is currently closed to the public per the City of Hazelwood.

17. *Shouldn't there be "caution" signage when sites are found?*
    The contamination that is being found is in subsurface soils and poses no risk in its current configuration. However, when the USACE finds contamination the property owner is informed and shown the location of the contamination. The USACE strongly suggests that the soil not be disturbed in these areas. If the property owner needs to disturb the soil where contamination exists, the USACE requests that the property owner contact us for support.

    Safety is our number one priority and proper posting is used as appropriate in accordance with local, state and federal regulations.

18. *What about the soil you have disturbed? What about dried soil/ dust?*
    The soil that has been disturbed is excavated soil and it is removed from the site. The soil is kept wet continuously to prevent dust emissions from the excavations. Air monitoring is conducted continuously during working hours when excavations are occurring.

19. *How is soil safe at subsurface when you are going to dig it up?*
    The contamination is located under clean soil which reduces the chances that the material will cause exposure. When it is dug up, the contamination is kept wet while being handled to prevent dust emissions that could cause exposure and then the soil is moved to the loadout area for disposal at a licensed facility.

20. *How are workers and residents protected from airborne soils?*
    Air monitoring is conducted continuously during working hours when excavation is occurring. While the excavations are being conducted, the soils are kept wet to prevent dust emissions. Soil that is removed from the excavation area is moved to the loadout area in covered trucks for eventual disposal.

21. *Doesn't ground water get affected by going through these areas?*
    Ground water monitoring is conducted to ensure ground water has not been contaminated. For more information on groundwater monitoring, see the Environmental Monitoring Data Analysis Reports (EMDARs) from 2014 on-line at www.mvs.usace.army.mil

22. *Cades Cove has a common area that often has standing water. Is that a higher risk?*

No, the standing water is not a higher risk. However, the builder used soils adjacent to CWC to fill low areas at Cades Cove to prevent flooding. The USACE is currently sampling in this area.

Risk is based on contamination levels and land use. The common area is being evaluated due to its low lying elevation.

23. *Describe "Safe Process" to transport, storage, clean up (trucks).*

The trucks used to carry contaminated soil to the load out facility are covered. The trucks are scanned before leaving an excavation area. If contamination is found the trucks are brushed off or hosed own with water and re-scanned to make sure no contamination is left.

See slides on the presentation.

24. *How does spray soil not allow contamination to seep further into soil?*

The contamination in the soil is made of particles just like the soil, so spraying water on excavation surface for the short time the excavation is being conducted does not cause the contamination to seep further into the soil. It just keeps the contaminated soil wet so that it does not become airborne.

25. *Who evaluates the results of the remediation? Agencies? Consultants? What studies do you rely upon to determine that contaminated soil no longer poses a risk?*

The North County Record of Decision (ROD) established the levels when remediation is needed. These levels are in compliance with CERCLA requirements. If sampling data is above the remedial goals determined by the ROD, then the area needs remediation. After remediation, the USACE and its contractors perform a Final Status Survey Evaluation (FSSE) which is reviewed by State and Federal regulatory agencies before the property is released for unrestricted use.

The FSSE demonstrates that the release criteria established by the Record of Decision (ROD) has been met. Demonstrating this requires collection of data for determining surface activity levels, direct exposure rates, and radionuclide concentrations.

26. *Is there current air monitoring? Is there tracking of potential windswept sites from long-uncovered piles of waste?*

Air monitoring is conducted at the North County sites. Results from air monitoring can be found in the Environmental Monitoring Data and Analysis Reports (EMDARs) which can be

found on-line at www.mvs.usace.army.mil. These documents
are published yearly with all the monitoring data (air, water,
and sediment) from the site.

27. *Is one elevated sample enough to remediate an area, or is this
    sample averaging across an area?*

    Our sampling plan is laid out in a grid format. If any of the grid
    samples have elevated levels we will return to do additional
    sampling to establish the limits of remediation requirements. It
    is possible for a single sample at a location to cause remediation
    to be required.

28. *Is the source of contamination gone?*

    Yes, the primary sources of contamination in North County
    have been cleaned up (St. Louis Airport Site (SLAPS) and the
    Latty Vicinity Properties (Hazelwood Interim Storage Site
    (HISS)/Futura and Latty properties).

29. *Can you tell me when the contamination happened based on soil
    depth?*

    No, there is no definite correlation between the depth of con-
    tamination and the date of contamination. However, in the case
    of CWC, since the contamination is subsurface, it would appear
    that the contamination is due to historic flooding in low-lying
    areas and the sediment was re-deposited in subsequent flood-
    ing events.

30. *What will happen to contaminated homes?*

    At this time, we have not found contamination on the exterior
    surfaces of any homes that we have investigated.

31. *Several times, CWC has overflowed its banks 30 feet from my
    front door. What is going to be done to keep that from happening
    again?*

    The FUSRAP mission is to remediate contamination remaining
    from Manhattan Engineer District / Atomic Energy Commis-
    sion processes. The Metropolitan Sewer District (MSD) is re-
    sponsible for controls at Coldwater Creek.

HOW DID THIS HAPPEN

32. *Can you determine what year the contamination started?*

    Approximately, yes. The earliest contamination would have
    been the result of runoff from the material stored at the SLAPS,
    which began in 1946.

There is an extensive history section on St. Louis FUSRAP website.

33. *Who is ultimately responsible and are they being held accountable?*

   The FUSRAP program was established by the Atomic Energy Commission (AEC) and later the Dept. of Energy (DOE) to clean up contamination from the atomic energy era. The US-ACE is currently responsible for cleanup of contamination in excess of the Record of Decision clean up criteria. Establishing responsibility for the contamination is not within the authority of the FUSRAP.

34. *Do you actually know where all 47,000 tons of illegally dumped waste from the Manhattan project actually is?*

   The FUSRAP is responsible for cleaning up the contaminated soil known to have been a result of atomic weapon development by the Manhattan Engineer District / Atomic Energy Commission and disposed of by their contractors, within the boundaries established in the Record of Decision.

35. *What is "Proper containment" for nuclear waste?*

   It would depend on where the waste is generated and on the properties of the waste itself. Proper containment would also depend on its current configuration and whether or not it has been disturbed, is in storage or is being shipped for disposal. Once contaminated soil is removed from the St. Louis FUSRAP sites, it is loaded into covered rail cars and shipped out of state to a licensed disposal facility.

36. *When did USACE become part of the clean up?*

   The USACE took over the execution of FUSRAP in 1997.

37. *How do you know this isn't part of Westlake?*

   SLAPS and HISS are adjacent to CWC and were the sources of contamination, not West Lake.

SAMPLING

38. *Will you be sampling properties in Jamestown?*

   We are sampling properties adjacent to Coldwater Creek.

39. *Will you contact property owners if their property is contaminated?*

   Yes. We personally meet with property owners when we find contamination on their property.

40. *Will you be sampling Berkeley properties?*

Many of the properties in the industrial area of FUSRAP are located in Berkeley. Many properties still need remediation and pre-design investigational sampling is ongoing south of Pershall Road which includes properties with the Cities of Berkeley and Hazelwood.

41. *How long before St. Cin Park was contaminated was it toxic?*

The contamination of St. Cin Park did not pose a risk because of its configuration below ground surface.

42. *I have heard it will be 2 years before mitigation efforts begin in Florissant. Why the delay and is this safe?*

We are currently investigating Coldwater Creek from upstream to downstream. We are sampling in areas south of St. Denis Bridge which are located in Florissant. We will start the next section of the creek from St. Denis Bridge to Old Halls Ferry Road late in 2016. Properties are added to the sites to be remediated as they are confirmed.

What were the findings from the samples taken at Normandie Court?

The investigation in this area is not yet complete. Once the investigation is complete, the USACE will provide information to property owners, as applicable. A Pre-Design Investigation Summary Report will also be completed and provided to property owners for review and comment prior to publishing.

43. *Will you be testing the Wedgewood Green Subdivision? There is a creek behind the subdivision. Is it possible the creek was rerouted and the soil was contaminated?*

This area is in the next section of the creek that will be investigated from St. Denis Bridge to Old Halls Ferry Road. Sampling will start late in 2016. To determine sampling locations, the USACE reviews historical data and aerial photographs to look for changes in topography, and evidence of being affected by a 10-year flood.

44. *Do you have any information about the McDonnell Blvd. area?*

McDonnell Blvd. was a main haul route and is also adjacent to the SLAPS. We have remediated the rights-of-ways (shoulder of the road) on McDonnell Blvd. adjacent to SLAPS. We have also found contamination under McDonnell Blvd. adjacent to the SLAPS. This contamination is located under the road and poses no risk in its current configuration. Land use controls will be placed on the contaminated soils under the road to protect the public and utility workers in the event that these soils need to be disturbed at a later time.

45. *Is the vegetation (trees, shrubs) along Coldwater Creek being tested? If so, do you have any results?*

    Vegetation testing is not part of the current plan being implemented by FUSRAP at Coldwater Creek; however, previous testing done by others has shown no evidence of contamination uptake by vegetation.

46. *Coldwater Creek runs behind my subdivision (Chapel Cross Drive). Was this area contaminated?*

    The area is located between St. Denis Bridge and Old Halls Ferry Road, which is the creek reach we anticipate beginning assessment of in late 2016.

47. *What is the risk of drinking the water from Coldwater creek (two of my dogs have had cancer since moving to Florissant in 2013)?*

    The USACE performs water monitoring from Coldwater Creek. We have not found MED/AEC contamination in the surface water. However, there are several other contaminants in the creek not related to the FUSRAP contaminants of concern.

48. *What is the risk to children under 5 playing at St. Ferdinand Park?*

    This park has not been investigated by the USACE. It is located in our next investigation area from St. Denis Bridge to Old Halls Ferry Road. However, the Missouri Dept. of Natural Resources has performed sampling in this park. You can contact them for results—Dan Carey 314-887-3046.

49. *Why are there no radioactive warning signs at St. Cin Park?*

    There is no need to post radioactive warning signs. The levels of radiation at St. Cin Park do not require "Radioactive" warning signs but we do use "Restricted Area" signage.

50. *Will you be sampling the creek on Humes in Flamingo Park?*

    Yes, it is in the next sampling area between St. Denis Bridge and Old Halls Ferry Road.

51. *Duchesne Park—When there is subsurface flooding due to rain, is it possible for contamination to come to the surface or into basements?*

Flooding of basements is likely caused by stormwater runoff backing up into homes from stormwater collection systems discharging to Coldwater Creek. We have found no contamination at the outside walls of homes. The configuration of the soil prevents contamination from leaching to the surface.

52.

53. *Do the watermarks represent flooded areas near Coldwater Creek?*
The USACE does not put watermarks for flooding. That would be a question for MSD who is responsible for the creek.

54. *Will tributaries to Coldwater Creek be sampled too?*
The mouths of the tributaries will be sampled initially, if contamination is found then we will continue to sample upstream in the tributary; as necessary.

55. *Will you be sampling/investigating Robertson, Missouri? Trains carried radioactive waste. History of cancer in family.*
Robertson was located west of the St. Louis Airport Site (SLAPS) and was part of the airport expansion. SLAPS has been remediated but the Robertson area is not within our FUS-RAP boundaries.

56. *Can you test the inside of my house? It's already been tested positive for uranium, thorium, and radium.*
If your property is within the 10 year floodplain, the soils on the property will be tested. This testing results will indicate whether testing inside the home should be considered. To date we have found no reason to suspect contaminants inside homes from soils that FUSRAP is remediating.

57. *Why is contamination showing up in certain areas and not all along the creek?*
Contamination is generally being detected at levels requiring remediation along the creek banks and low lying, or formerly low lying, properties adjacent to the creek. Properties above the 10-year floodplain typically are not affected.

58. *Have properties along the creek all been tested?*
Properties adjacent to Coldwater Creek from Banshee Road to just south of St. Denis Bridge have been tested. We anticipate beginning testing north of St. Denis Bridge in late 2016.

59. *Has there been testing on the property where the new Wal-Mart is located?*

Yes, the USACE performed some sampling and testing in areas adjacent to the creek during the building of the Wal-Mart in Florissant. No contamination was found.

60. *When was the last nuclear material removed from McDonnell Douglas? Is there any remaining contamination at these buildings?*

The property surrounding the Boeing Buildings was sampled. No Manhattan Engineer District/Atomic Energy Commission (MED/AEC) contamination was found. There was no reason to enter the buildings at Boeing to investigate MED/AEC contamination.

61. *Has sampling been performed where Missouri American water lines cross Coldwater Creek in Florissant (ex: Washington Street Bridge)?*

We are still sampling that area.

62. *What is the status of sampling on Coldwater Creek, the tributary from B&K's property, St. Ann Park, and the airport where B&K washed off their vehicles?*

This area is not within the FUSRAP boundaries defined in the North County Record of Decision.

63. *When is the area near Hazelwood Central High School going to be sampled?*

The USACE is sampling properties within the 10 year floodplain adjacent to Coldwater Creek from Banshee Road to the Missouri River. It is anticipated that sampling of the section of Coldwater Creek between St. Denis Bridge and Old Halls Ferry Road, where Hazelwood Central High School is located, will begin late in 2016.

64. *Will you be sampling in the Hathaway Manor Subdivision?*

The USACE is sampling properties within the 10 year floodplain adjacent to Coldwater Creek from Banshee Road to the Missouri River. It is anticipated that sampling of the section of Coldwater Creek between St. Denis Bridge and Old Halls Ferry Road will begin in late 2016.

65. *Have all the original drums of radioactive material been removed?*

Yes, the original drums have been removed and the SLAPS site has been remediated. Remediation at this property was completed in 2007.

66. *Will you sample around Maline Creek / Northland Hills Subdivision?*

No. Maline Creek is not within the FUSRAP boundaries and it does not connect with Coldwater Creek.

67. *Will there be sampling at St. Ferdinand Park and the Knights of Columbus Park?*

We are currently sampling Knights of Columbus Park. St. Ferdinand Park will be sampled in the next sampling campaign from St. Denis Bridge to Old Halls Ferry Road.

St. Ferdinand Park was tested by Missouri Department of Natural Resources. No contamination was found.

68. *Is the creek between Old Hall Ferry Road and New Halls Ferry Road going to be sampled?*

Yes, this section of creek will be sampled during the next sampling campaign from St. Denis Bridge to Old Halls Ferry Road.

69. *Will you be sampling homeowner's yards?*

We plan to sample yards that are adjacent to Coldwater Creek within the 10 year floodplain.

70. *Does Coldwater Creek run off into other connecting creeks and rivers?*

Many tributaries feed into Coldwater Creek which flows into the Missouri River.

71. *Why was contamination found 18 inches below the ground and not on top of the ground?*

Contamination has been covered over the years by other sediments and soils from flood water and land development activities.

72. *Will residences along Patterson be sampled?*

Only if adjacent to Coldwater Creek and within the 10 year flood plain.

73. *How is the 10 year flood plain determined"*

The Pre-Design Investigation Work Plan for CWC from Frost to St. Denis Bridge included information for Coldwater Creek used to find elevations at known locations with the 10% chance to flood in a given year. The 10 yr. floodplain was then determined by extracting and connecting corresponding elevations from current elevation model data using a Geospatial Information System (GIS).

74. *Can you please explain the significance of the 10 year flood plain as the sampling limit? Is there any reason to expect contamination beyond that limit?*

The 10 year flood plain is not the sampling limit but the sampling beginning. We are starting with the 10 year flood plain and will investigate further beyond that area if contamination above Record of Decision clean up goals are found on the outer edge of the 10 year flood plain.

75. *Can we ensure that sampling is being performed beyond the flood plain?*

See previous comment.

76. *What was found at Duchesne Park? What are the risks of playing/visiting the park?*

Low levels of contamination were found at the park. The contamination found poses no risk in its current configuration below ground surface.

77. *Will South St. Charles Street be sampled?*

Only if adjacent to Coldwater Creek and within the 10 year floodplain.

79.

80. *Who notifies renters of contamination?*

The USACE will notify renters and property owners of contaminated areas if contamination is found.

77. *Can you explain the impacts of Coldwater Creek's confluence with the Missouri River?*

We have not sampled this area yet and cannot give you any answers about contamination from Coldwater Creek into the Missouri River. This is the last stretch for sampling. Naturally you would expect contamination to decrease as you move away from the source (SLAPS and HISS).

78. *Is the area between Hanley and Latty in Hazelwood affected? Should I be concerned about gardening / eating vegetables planted in this area?*

This area is not within the FUSRAP boundaries. Previous studies performed on vegetation near CWC have not revealed any contaminant uptake by vegetation.

79. *Should I be concerned if I ate vegetables from a contaminated area 20 years ago?*

There has been no evidence of an uptake of contamination into vegetation.

80. *What areas have been tested?*

    We have tested the areas within the FUSRAP industrial area south of Hwy 270. The areas north of Hwy 270, we are currently sampling to St. Denis Bridge. We will start the next sampling campaign from St. Denis Bridge to Old Halls Ferry Road in late 2016.

81. *How can I request an area to be tested?*

    We cannot sample a property by request. We can only sample areas within the FUSRAP boundary established by the North County Record of Decision. If the property owner has specific reason to discuss the likelihood of contamination on their property, they can contact USACE at 314-260-3905.

82. *Can you discuss testing for U-235 and Thorium? Where are results from sampling located?*

    While the contaminants at the North St. Louis County Site do not include U-235, samples are analyzed for U-235 and thorium. The results will be posted on the website once the sampling and analysis is completed to St. Denis Bridge and the document is released.

83. *At what level above background have the contaminated sites been tested?*

    Testing has found of range of approximately 1–20 times background levels, on average. It should be noted that of all the samples collected in and along the CWC (approximately 8500) less than 5% of these locations tested positive for contamination.

84. *Does Coldwater Creek run through the Pleasant Hollow subdivision?*

    No, Pleasant Hollow subdivision is approximately 1.5 miles west of Coldwater Creek.

85. *Should children play in St. Ferdinand Park?*

    This park has not been investigated by the USACE. It is located in our next investigation area from St. Denis Bridge to Old Halls Ferry Road. However, the Missouri Dept. of Natural Resources has performed sampling in this park. You can contact them for results—314-887-3046.

86. *Will Chez Paree be sampled?*

Chez Paree has been sampled and is part of the Palm Drive Properties that will be remediated when all pre-remediation activities are complete. Contamination was found in the back of the complex adjacent to Coldwater Creek.

87. *How deep is the contamination at Duchesne and Palm Drive?*
   At Duchesne Park, contamination was found from .5 feet down to approximately 1.5 feet below ground surface. At Palm Drive, the contamination was found from .5 feet down to approximately 2 feet below ground surface.

88. *Microbes can percolate to the surface of the ground, is that a threat?*
   FUSRAP addresses radioactive contamination only.

WATER

90. *What tests were completed on ground water contamination?*
   Groundwater is sampled for the FUSRAP Contaminants of Concern and the results are contained in the Environmental Monitoring Data and Analysis Reports (EMDARs), which are posted on the FUSRAP website.

91. *What were the results from ground water contamination testing?*
   Results can be found in the Environmental Monitoring Data and Analysis Reports, which are posted on the FUSRAP website.

92. *Why are only 3 properties on Palm Drive contaminated?*
   Sampling/investigating the properties on Palm Drive showed the backyards of four properties adjacent to Coldwater Creek that were contaminated. Contamination at Palm Drive was possibly caused by flooding. The USACE sampled other properties on Palm Drive that were also flooded but no contamination was found.

93. *Home backs up to Coldwater Creek and has flooded several times, is this from Coldwater Creek or something else (MSD backups?)?*
   Depending upon your home location and the nature of the flooding referenced (i.e., backyard versus basement) the water could be from either source.

94. *Does the creek affect drinking water / water supply?*

No, the creek does not affect drinking water.

95. *Is drinking water tested for radionuclides?*
    FUSRAP does not address drinking water. You will have to contact Missouri American Water to answer that question.

96. *If the water comes into your house or you are located directly behind the creek are you at risk?*
    Surface water testing results, which can be found in the annual Environmental Monitoring Data and Analysis Reports on the FUSRAP website, do not indicate FUSRAP COCs in the water within Coldwater Creek.

97. *Does the Florissant Government know about the contamination?*
    Yes, the USACE is working with the City of Florissant.

98. *Do you test all of the ground water prior to remediation?*
    Yes, we monitor groundwater before, during, and after remedial activities. Groundwater results can be found on-line in the Environmental Monitoring Data and Analysis Reports at www.mvs.usace.army.mil.

SOILS THAT MOVED

99. *Are there plans to sample soils of the areas that were subsequently built over the creek after it moved from its natural location?*
    The USACE studies all the historical aerial photographs and other historical background information regarding the creek to determine sample locations, including areas that have been affected by development.

100. *During a 1997–1999 Coldwater Creek erosion project, the banks were excavated. What impact might that have?*
    The USACE was not part of the erosion project. We will still sample the banks through the concrete and rip-rap on the banks to determine if contamination is located in these areas. If you have information on the project, please send it to us so we can evaluate it 314-260-3905.

101. *Was the dirt for the flood plain behind Cades Cove subdivision used to level the ground where the house was placed?*
    It is our understanding that soil near the bank of Coldwater Creek was used to elevate the back yards at Cades Cove. Given

this vital information, the USACE is currently sampling the soils in this area even though they are outside the 10 year flood plain.

102. *It was stated several times that areas are not dangerous in their current configuration. Were these areas more dangerous in a previous configuration?*
Current configuration refers to the contaminants being found .5 feet to 2.5 feet below surface, depending on location; the contamination is low level radioactive contamination. It is hard to tell when the contaminants were deposited and by what mechanism, and therefore what level of risk. Levels that are being found currently present no risk.

103. *Were areas that are now a foot or two deep closer to or on the surface 30 years ago?*
Possibly, due to the accumulation of sediment over time, or movement of soil by developers.

REMOVAL AND REMEDIATION

104. *What is the level that determines "Clean," since acceptable levels have changed over time?*
The USACE follows the North County Record of Decision health-based remedial goals that establish "clean."

105. *What is the acceptable percentage of recovery/remediation at a site?*
100% of the contaminated soils that exceed the Record of Decision cleanup goals are removed.

106. *After a site is determined to be "clean," what is follow-up sampling schedule to ensure the clean-up is permanent?*
The USACE follows the North County Record of decision health-based remedial goals that establishes "clean." The area of excavation is sampled after the completion of remediation and if determined "clean" it is backfilled with clean backfill material. There is no follow up sampling after the area has been verified to be "clean."

107. *What does "extensive preparatory work" consist of?*
The extensive preparatory work for the Palm Drive properties consists of re-locating utility poles, removing trees, re-locating back yard fences and building a haul road.

108. *Is my garden safe to eat from? Was it 10 years ago?*
Garden vegetables have not been tested as part of the FUSRAP
project; however, previous testing of trees and shrubs by others
near Coldwater Creek has shown no evidence of contamination
uptake by vegetation.

PROPERTY VALUE

109. *Is this now a required part of Sale disclosures?*
The USACE is unable to answer that question. You will need to
consult with a real estate specialist.

110. *Is there a buyout?*
No, the Record of Decision remedy is for excavation and ship-
ping/disposal of contaminated soils. It does not include a buyout option.

111. *My appraisal is dropping, but my assessment isn't. How will I
be compensated?*
The FUSRAP Record of Decision remedy does not include re-
imbursement for drop in fair market value of homes.

TIME/FUNDING/LENGTH OF PROJECT

112. *Who do I need to contact to influence the urgency of this
situation?*
The USACE realizes the urgency of the situation. We are working as fast as possible. Sampling takes time to thoroughly investigate the potential areas of contamination. There are more
than 14 miles of creek and adjacent properties to investigate.

113. *How can I raise money for this cause? When I do, who should I
give it to?*
The FUSRAP is funded directly by Congress. We cannot accept
donations.

114. *Why do you only work in one area at a time if you have lots of
funding?*
The St. Louis FUSRAP Program manages five sites out of the
25 FUSRAP sites in the FUSRAP Program throughout the nation. We have to share this funding provided by Congress with
the rest of the FUSRAP sites. The St. Louis FUSRAP Program
consists of the St. Louis Downtown Site (SLDS), the St. Louis
Airport Site (SLAPS), the Latty Avenue Sites (which includes

HISS/Futura and adjacent properties), the SLAPS Vicinity Properties (which includes Coldwater Creek and currently over 148 properties) and the Iowa Army Ammunition Plant (IAAP) which all need some level of funding for the various phases of remediation. Currently, remediation and/or sampling is occurring along Coldwater Creek, at the SLDS site and in Iowa.

115. *What is the status of the procurement and selection of new US-ACE FUSRAP contractor? Is there an award date?*

   USACE is currently evaluating contractor proposals for the remedial action contract.

116. *Has sediment sampling occurred yet?*

   Yes, sediment sampling is ongoing. Results of yearly sediment sampling can be found in the Environmental Monitoring Data and Analysis Reports on the St. Louis FUSRAP website.

117. *How would USACE remove any contaminated sediments?*

   Likely that methods currently being used would also be used in the creek. There will be continual sampling and monitoring of creek waters. The creek sediment will be excavated the same as soil. It will be excavated, shipped and disposed in an out of state licensed facility.

   Why not spread testing out so we don't have to wait for years to find out results?

   The USACE is working from upstream to downstream to locate areas of contamination in and adjacent to the creek. We need to concentrate on one area at a time to complete sampling and prepare to excavate if necessary. Property owners are notified immediately if contamination requiring remediation is found on their property.

118. *How long does it take to complete the phases? Any recommendations for people who live in future phases?*

   There is no set time to complete the phases. Sampling depends on various factors such as weather conditions; obtaining rights of entry; flooding; additional areas to sample; etc.

119. *Did the government not think the barrels would rust and leak when they disposed of them?*

   We cannot make assumptions and speculate on decision making actions surrounding historic events.

120. *Where has the money that has been allotted to this program for years gone?*

The USACE continues to remediate the St. Louis Downtown Site (SLDS) and has shipped approximately 274,000 cubic yards of contaminated materials; the SLAPS site was completed with over 600,000 cubic yards of contaminated materials shipped; the HISS/Futura/Latty properties are completed with over 250,000 cubic yards of contaminated materials shipped. Many of the vicinity properties in North County have been remediated to date (over 65,000 cubic yards of contaminate materials shipped). Also, the St. Louis FUSRAP Program includes the Iowa Army Ammunition Plant where we are performing remediation; St. Louis FUSRAP completed the Madison, Ill., site in 2000, and we are currently investigating/remediating Coldwater Creek To ensure the maximum protection and safety of the public, the processes used are necessarily detailed and risk-reduction based.

121. *What is the timetable for work on Phase 2?*
We anticipate sampling from St. Denis Bridge to Old Halls Ferry Road will start in late 2016.

122. *What can we do to speed up the cleanup process? Is there anything we can do?*
The USACE requests that property/business owners who receive a right-of-entry (ROE), sign the ROE and send it back as soon as possible. If you have any questions of the ROE, please call and we will set up a meeting to discuss your concerns.

HEALTH QUESTIONS

The FUSRAP mission is to address releases or threatened release of hazardous substance related to the nation's early atomic energy program that post an unacceptable risk to human health and the environment. USACE employees are not medical experts and do not have the expertise to offer any opinions about the cause of any illness in the area.

For any medical related questions please contact Erin Evans for the Agency for Toxic Substances and Disease Registry (ATSDR) at evans. erin@epa.gov OR the St. Louis County Department of Public Health.

WESTLAKE LANDFILL

The Westlake Landfill is not a FUSRAP project.

COLDWATER CREEK TIMELINE

1831 German immigrant Emil Mallinckrodt purchases a plot of land north of St. Louis.

1867 Emil Mallinckrodt's three sons found G. Mallinckrodt and Company on their family farm. As the only chemical company west of Philadelphia, it captures western markets.

1876 City voters decide to separate from St. Louis County, making the City of St. Louis landlocked within a sixty-one-square-mile area. The separation becomes final the following year.

1880 Concerned about industrial pollution, the St. Louis board of health adopts a policy of selective law enforcement that prioritizes property values over the health of residents.

1882 Mallinckrodt Chemical Works is incorporated.

1896 Albert Bond Lambert leaves the University of Virginia to become president of his family's business.

1910 Lambert's Aero Club of St. Louis establishes the "Permanent Aviation Field and Dirigible Harbor," popularly known as the Kinloch Flying Field.

1920 Albert Bond Lambert and the Missouri Aeronautical Society obtain a five-year lease (with option to purchase) on 170 acres of farmland in Bridgeton, Missouri.

1923 The facility at Bridgeton is dedicated as the St. Louis Flying Field.

1925 Lambert purchases the St. Louis Flying Field.

1927 Lambert offers the St. Louis Flying Field to the City of St. Louis as one of the first municipal airports in the United States. The city's acceptance is contingent on the passage of a bond issue.

1928 Edward Mallinckrodt Jr. assumes the leadership in his family-owned business.

1930 Lambert–St. Louis Airport is dedicated. To facilitate runway construction, the City of St. Louis authorizes $275,000 to reroute Coldwater Creek, the first of many airport alterations to the waterway.

1941 The United States enters World War II.

1942 City voters pass a bond issue for airport expansion.

The Army Corps of Engineers establishes the Manhattan Engineer District with a secret group (the Manhattan Project) focused on the development of the first atomic bomb.

Mallinckrodt Chemical Works begins uranium processing as part of the Manhattan Project.

Enrico Fermi directs the first controlled, self-sustaining nuclear chain reaction with uranium purified at Mallinckrodt Chemical Works.

1945 The United States drops an atomic bomb on Hiroshima. It is made of uranium processed at the Mallinckrodt Chemical Works.

World War II ends.

1946 The Manhattan Engineer District obtains consent to use 21.7 acres adjoining the airport for storing the radioactive process waste from the first atomic bomb. Called the St. Louis Airport Storage Site (SLAPS), this property is bordered by Coldwater Creek. Even before the land is acquired by condemnation, trucks carry the waste to the property.

City voters pass another bond issue for airport expansion. The project requires the full enclosure of Coldwater Creek on airport property.

1947 to mid-1970s   North St. Louis County undergoes a housing boom. Subdivision development requires the reconfiguring of Coldwater Creek. Radioactive contamination spreads from SLAPS to residential areas via the creek.

1966 Continental Mining and Milling purchases some residues and wastes stored at SLAPS and begins moving it one-half mile to 9200 Latty Avenue. During the transport many properties are contaminated from spilled radioactive waste.

1967 The Atomic Energy Commission (AEC) authorizes the use of SLAPS by the City of St. Louis.

1969 Cotter Corporation acquires the process waste from SLAPS and begins shipping it to the company's plant in Colorado. The process leaves 8,700 tons of leached barium sulfate at the Latty Avenue site. The barium, used to recover uranium, contains uranium residue.

1973 Cotter mixes the barium sulfate with thirty-nine-thousand tons of topsoil and contracts with B&K Construction Company to ship it to West Lake Landfill in nearby Bridgeton, Missouri.

The St. Louis Airport Storage Site is transferred from the federal government to the city by quitclaim deed.

1974 The AEC notifies Cotter Corporation that the disposed material at

West Lake is radioactive and in violation of federal regulations. The AEC takes no enforcement action against the company. Its successor, the National Regulatory Commission (NRC), does not exercise its legal right to request that the material be retrieved and placed in suitable storage.

The AEC establishes the Formerly Utilized Sites Remedial Action Program (FUSRAP) to remediate or control sites where the Manhattan Engineer District or the AEC engaged in activities from the 1940s through the 1960s. The downtown Mallinckrodt plant, St. Louis Airport Storage Site, and Latty Avenue properties are included, along with seventy-eight vicinity properties in Hazelwood and Berkeley, Missouri.

1976 A survey for the U.S. Department of Energy finds some erosion from SLAPS into Coldwater Creek. Sediment samples from the creek bed show levels of radium-226 that are ten times above natural background levels.

1977 Dean Jarboe purchases 3.5 acres in the 9200 block of Latty Avenue to build a headquarters for his company, Futura Coatings. Officials inform him that he cannot build because the property is contaminated.

1980 Jarboe purchases an additional 7 acres to serve as interim storage for the contaminated and demolished building rubble cleared from his 3.5-acre lot. He has plans to expand his operation after the federal government removes the radioactive material. However, the consolidated waste (known as the Hazelwood Interim Storage Site) is still awaiting removal in 1996.

1984 The U.S. Congress directs the Department of Energy (DOE) to reacquire SLAPS from the City of St. Louis and study options for disposing of it and the Latty Avenue wastes. The city does not immediately transfer the property back to the DOE as authorized by law.

1985 As contaminated soil slowly erodes into Coldwater Creek, the DOE builds a gabion wall on the creek bank to try to contain it.

1986 The Superfund Amendments and Reauthorization Act refines its risk calculation for assessing hazardous wastes.

1988 The city board of aldermen's Special Committee on Radioactive Waste issues a report urging the Missouri congressional delegation to direct the DOE to find an environmentally sound disposal site for St. Louis wastes—far away from a population center. This position is supported by a nonbinding referendum in which voters in St. Louis City and County overwhelmingly oppose constructing a radioactive waste bunker at the SLAPS site or any location near the City of St. Louis.

The NRC issues a report describing radioactive contamination at West Lake Landfill in Bridgeton.

1989 EPA places both the SLAPS and Latty properties on the Superfund National Priorities List.

1990 At the request of the Missouri Department of Natural Resources (MDNR), EPA includes the West Lake Landfill as a Superfund site.

Congress passes the Environmental Education Act.

1992 EPA initiates Superfund enforcement action against potentially responsible parties at West Lake Landfill. They include the DOE and the Cotter Corporation.

1993 EPA issues a consent order against potentially responsible parties at West Lake, requiring them to conduct a remedial investigation / feasibility study on the site.

1994 The St. Louis Site Remediation Task Force is appointed by DOE and charged with identifying and evaluating remedial-action alternatives for St. Louis FUSRAP sites and the West Lake Landfill. The task force is also responsible for petitioning DOE to undertake a cleanup strategy that is environmentally acceptable and responsive to health and safety concerns.

1995 The DOE begins shipping radioactive waste from the St. Louis area to Utah to a disposal facility run by Envirocare.

1996 The St. Louis Site Remediation Task Force identifies SLAPS as the first cleanup priority, and thereafter (in no special order) the Ballfield Site in Berkeley, North City and North County Haul Roads, HISS / Futura Coatings, and Coldwater Creek. Task force members report insufficient data to judge the long-term health and environmental effects of contaminants at the creek. They recommend a cleanup in keeping with the highest standards. A panel of geologists and hydrologists finds the SLAPS contaminants at Coldwater Creek are a chronic problem but not acute.

1998 SLAPS responsibility shifts from DOE to the Army Corps of Engineers.

2011 Concerned citizens organize in a Facebook group called "Coldwater Creek—Just the Facts Please."

2013 The Coldwater Creek Facts group gathers data in an online health survey for residents. The group also discovers that the Army Corps of Engineers has confined its cleanup work to areas around SLAPS and Latty Avenue.

The Missouri Department of Health and Human Services releases *Analysis of Cancer Incidence Data in Coldwater Creek Area, Missouri, 1996–2004*. It finds an increased risk to cancer exposure unlikely.

Diane Whitmore Schanzenbach cites faulty methodology and disputes the state's findings in an article.

2014 The Missouri Department of Health and Human Services revises its study to account for Schanzenbach's criticisms. The new study finds higher-than-normal rates of some cancers.

In response to citizen concerns, the Army Corps of Engineers begins testing for the first time in areas north of Interstate 270.

2015 The corps discovers radioactive contamination in twelve sites along Coldwater Creek. It is the first time for the verification of contamination in residential properties.

The Centers for Disease Control (CDC) announces that scientists from its Agency for Toxic Substance Disease Registry (ATSDR) will team up with county health officials to assess health risks in and around Coldwater Creek.

2018 The ATSDR report finds that children and adults who regularly played in or around Coldwater Creek or lived in its floodplain for many years in the past (1960s to 1990s) may have been exposed to radiological contaminants and have an increased chance of developing certain cancers.

GLOSSARY OF TERMS

Agency for Toxic Substances and Disease Registry (ATSDR): A federal public-health agency of the U.S. Department of Health and Human Services. It protects communities from harmful health effects related to exposure to hazardous substances.

alpha radiation: An emission of a nucleus of high kinetic energy from the nucleus of an atom undergoing radioactive decay.

aquifer: A body of permeable rock that can contain or transmit groundwater.

Army Corps of Engineers (ACE): A federal agency under the Department of Defense that primarily oversees dams, canals, and flood-protection projects in the United States. It also handles environmental restorations.

atomic bomb: Weapon with great explosive power that results from the sudden release of energy upon the splitting, or fission, of the nuclei of a heavy element such as uranium or plutonium.

atomic chain reaction: See "nuclear chain reaction (self-sustained)."

Atomic Energy Act: A federal law covering the development, regulation, and disposal of nuclear materials and facilities in the United States.

Atomic Energy Commission (AEC): Agency of the U.S. government created in 1946 to manage the development, use, and control of atomic (nuclear) energy for military and civilian applications.

autoimmune: Relating to a condition in which someone's antibodies attack substances that are naturally found in the body.

background level: The concentration of a hazardous substance that provides a defensible reference point that can be used to evaluate whether a release from the site has occurred.

barium sulfate: An inorganic compound, this white crystalline solid is odorless and insoluble in water.

biosphere: The part of the earth and its atmosphere in which living things exist.

cancer cluster: A greater-than-expected number of cancer cases among a group of people in a defined geographic area over a specific time period.

Centers for Disease Control (CDC): The leading national public-health institute of the United States.

chinook: A warm, dry wind that blows down the east side of the Rocky Mountains at the end of winter.

confluence: A coming or flowing together at one point.

consent order: Generally a voluntary agreement worked out between two or more parties in a dispute. It usually has the same effect as a court order and can be enforced by the court if anyone does not comply with the order.

crawdad: A slang term for "crayfish."

creek: A natural stream of water normally smaller than and often a tributary to a river.

culvert: A tunnel carrying a stream or open drain under a road or railroad.

curie: One of three units used to measure the intensity of radioactivity in a sample of material. This value refers to the amount of ionizing radiation released when an element (such as uranium) spontaneously emits energy as a result of the radioactive decay of an unstable atom.

Department of Energy (DOE): A cabinet-level department of the U.S. government concerned with the nation's policies on energy, environment, and nuclear challenge.

eminent domain: The right of a government or its agent to expropriate private property for public use, with compensation to private owners.

Energy Research and Development Administration (ERDA): A former U.S. government agency formed from the split of the Atomic Energy Commission in 1975.

enriched uranium: A type of uranium in which the percentage composition of uranium-235 has been increased through the process of isotope separation.

Environmental Education Act (1990): An act of the U.S. Congress that provides, through the Environmental Protection Agency, resources to local institutions and not-for-profit educational and environmental organizations to support and improve awareness of environmental problems.

Environmental Justice Watershed: An EPA designation based on an area's percentage of low-income and minority residents. The program entitles recipients to funding and technical assistance to address environmental and public-health issues.

Environmental Protection Agency (EPA): An agency of the U.S. government with a mission to protect human and environmental health.

Federal Emergency Management Agency (FEMA): A federal agency with a primary purpose of coordinating the response to a disaster in the United States that overwhelms the resources of federal and state authorities.

Federal Facilities Compliance Act: Amended the Solid Waste Disposal Act to waive the sovereign immunity of the United States for purposes of enforcing federal, state, interstate, and local requirements regarding solid- and hazardous-waste management.

feed materials: Refined uranium or thorium metal or their pure compounds in a form suitable for use in nuclear-reactor fuel elements, or as food for uranium-enrichment processes.

fissile: Capable of sustaining a nuclear-fission chain reaction.

fission: The action of dividing or splitting something into two or more parts. Nuclear fission is a reaction in which a heavy nucleus splits spontaneously or on impact with another particle, releasing energy.

floodplain: An area of low-lying ground adjacent to a river, formed mainly of river sediment, and subject to flooding.

Formerly Utilized Sites Remedial Action Program (FUSRAP): A program initiated in 1974 to identify, investigate, and clean up or control sites throughout the United States contaminated as a result of the nation's early atomic weapons and energy programs.

gabion wall: A retaining wall made of stacked, stone-filled containers connected with wire.

groundwater: Water held underground. It is stored in and moves through geologic formations of soil, sand, and rocks called aquifers.

half-life: The time for the radioactivity of a specified isotope to fall to half its original value.

Hanford: A small agricultural community in Washington State that was depopulated in 1943 to make room for the nuclear-production facility known as the Hanford Site.

herbivorous: Eating plants.

hydrosphere: The water environments of the earth.

ingrowth: A thing that has grown inward or within something.

iodine-131: A radioisotope of iodine with a radioactive decay half-life of about eight days.

ionizing radiation: A radiation with enough energy so that during an interaction with an atom, it can remove tightly bound electrons (negative particles) from the orbit of an atom, causing the atom to become charged or ionized.

isotope: Each of two or more forms of the same element that contain equal numbers of protons but different numbers of neutrons in their nuclei, and therefore differ in relative atomic mass but not in chemical properties; in particular, "isotope" refers to a radioactive form of an element.

leach: Remove a chemical substance by the action of water passing through the material.

leukemia: A type of cancer that affects the production and function of blood cells.

loessial soils: Predominantly silt-sized sediment formed by the accumulation of windblown dust.

Los Alamos: A town in New Mexico that is recognized as the development and creation place of the atomic bomb.

lupus: A systemic autoimmune disease that occurs when people's immune systems attack their own tissues and organs.

lymphoma: A cancer that begins in infection-fighting cells of the immune system called lymphocytes. These cells are in the lymph nodes, spleen, thymus, bone marrow, and other parts of the body. When a person has lymphoma, lymphocytes change and grow out of control.

Manhattan Engineer District (MED): The code name for the army's involvement in the Manhattan Project.

Manhattan Project: The research and development undertaking in World War II that produced the first nuclear weapons.

maximum permissible dose: An amount of radiation exposure that, in light of present knowledge, is not expected to cause appreciable bodily injury.

Missouri Department of Natural Resources (MDNR): State agency with a mission to protect air, land, and water; preserve natural and historic places; and provide recreational and learning opportunities for Missourians.

National Cancer Institute: The U.S. government's principal agency for cancer research and training.

National Committee on Radiation Protection (NCRP): Established in 1929 as the Advisory Committee on X-Ray and Radiation Protection, it has provided basic standards and guidance in the field.

National Defense Research Committee (NDRC): An organization that began research on what would become some of the most important technology in World War II, including radar and the atomic bomb. It was founded in 1940 and superseded by the Office of Scientific Research and Development in 1941. The NDRC then became an advisory group until being terminated in 1947.

National Environmental Policy Act: The first major environmental policy law in the United States, signed in 1970.

nuclear chain reaction (self-sustained): The process by which neutrons released in the splitting of an atomic nucleus produce an additional splitting in at least one further atomic nucleus. This nucleus in turn produces neutrons, and the process repeats.

Nuclear Regulatory Commission (NRC): Independent government agency charged with overseeing civilian use of nuclear energy.

nucleus: The positively charged central core of an atom, consisting of protons and neutrons and containing nearly all its mass.

Oak Ridge, Tennessee: An Appalachian town that was the home of uranium enrichment during World War II.

Office of Scientific Research and Development (OSRD): A federal agency created to coordinate scientific research for military purposes during World War II.

picocurie: One-trillionth of a curie, which is a unit used to measure the intensity of radioactivity in a sample of material. This value refers to the amount of ionizing radiation released when an element (such as uranium) spontaneously emits energy as a result of the radioactive decay of an unstable atom.

piscivorous: Feeding on fish.

pitchblende: An ore of the mineral uraninite occurring in brown and black masses and containing uranium.

plutonium: A radioactive chemical element that is formed in nuclear-power reactors from uranium-238 by neutron capture. All fifteen of its isotopes are radioactive.

plutonium pit: The radioactive core of a modern nuclear weapon. It was essentially a plutonium pit that destroyed Nagasaki.

plutonium-239: An artificial radioactive nucleus produced in large quantities by reactors when nuclei of uranium-238 capture an extra neutron apiece.

potable: Drinkable.

prairie: A large, open grassland, especially in the Mississippi River valley.

process waste: Pollutant or combination of pollutants that are designated as toxic, present in wastewater or inherent to a manufacturing or production process, and discharged as waste into the environment.

quitclaim deed: A legal instrument used to transfer interest in real property.

radiation: The emission of energy as moving subatomic particles, especially high-energy particles that cause ionization.

radioactive: A description of isotopes that are unstable in that they have an imbalance of neutrons and protons. They release this energy over time and undergo radioactive decay to a stable isotope.

radionuclide: An atom that has excess nuclear energy making it unstable.

radium: A naturally occurring radioactive element. It is a radionuclide formed by the decay of uranium and thorium in the environment.

radium-226: A radioactive heavy metal that is one of the more dangerous of the uranium-decay products. As it decays, it produces radon gas as a byproduct.

radium-236: An isotope of radium that is neither fissile (able to undergo nuclear fission) with thermal neutrons nor a very good fertile material, but is generally considered a nuisance and long-lived radioactive waste. It is found in spent nuclear fuel and in reprocessed uranium made from spent nuclear fuel.

raffinate: A liquid from which impurities have been removed by solvent extraction.

residue: A small amount of something that remains after the main part has gone or been taken away or used.

sediment: The material from a liquid that settles to the bottom of the liquid's container.

sedimentation: The process of settling or being deposited as a sediment.

sink: (1) A topographic depression formed when an underlying limestone bedrock is dissolved by groundwater; (2) a place of deposit, according to environmental historian Joel Tarr.

source materials: Material containing either the element thorium or the element uranium, provided that the uranium is not enriched in the element uranium-235 above a level found in nature.

St. Louis Airport Storage Site (SLAPS): A 21.7-acre tract used to store residues from uranium processing from the Mallinckrodt Chemical Works.

St. Louis Site Remediation Task Force: A broad-based representative body formed in 1994 to identify and evaluate remedial-action alternatives for the cleanup and disposal of radioactive waste materials at the St. Louis FUSRAP site and the West Lake Landfill.

strontium-90: A radioactive isotope of strontium produced by nuclear fission, with a half-life of 28.8 years.

Superfund Amendments and Reauthorization Act: Passed in 1986, this act amends the Comprehensive Environmental Response, Compensation, and Liability Act (CERCLA) to help solve the problems of hazardous-waste sites.

Superfund National Priority List: The list of sites of national priority among the known releases or threatened releases of hazardous substances, pollutants, or contaminants throughout the United States and its territories.

thorium: A naturally occurring, slightly radioactive metal discovered in 1828 by Swedish chemist Jon Jakob Berzelius. It is much more abundant on Earth than uranium and is fueled primarily by the nuclear fission of uranium-233.

thorium-230: An isotope of thorium. Thorium-230 is a decay product of uranium-238.

thorium-232: This isotope of thorium is unstable. As it decays it releases radiation and forms decay products that include radium-228 and its decay product thorium-228 (which then decays into radium-224).

tolerance dose: The largest quantity of an agent that may be administered without harm.

topsoil: The top layer of soil.

toxic: Acting as or having the effect of a poison.

tributary: A stream feeding a larger steam, river, or lake.

uranium: A heavy metal that can be used as an abundant source of energy. Its symbol is "U," and its atomic number is ninety-two.

uranium processing: One step in nuclear weapons production, occurring after uranium mining and before uranium enrichment.

uranium refining: The process by which dried uranium (in a powder called yellowcake) goes through a procedure to become three products: uranium oxide, natural metallic uranium, and uranium hexafluoride.

uranium-234: An isotope of uranium. In natural uranium and in uranium ore, U-234 occurs as an indirect decay product of U-238.

uranium-235: An isotope of uranium. Unlike the predominant isotope U-238, it can withstand a nuclear chain reaction.

uranium-238: The most common isotope of uranium found in nature. Unlike U-235, it cannot withstand a nuclear reaction.

watershed: An area of land that drains to a stream, lake, or wetland.

NOTES

Introduction

1. Ionizing radiation is a flow of energy capable of freeing electrons from an atom, causing it to be charged. The process can cause extensive damage to cells and DNA, resulting in cancer and even death.

2. Tarr, *Search for the Ultimate Sink*.

3. Brown, *Plutopia*.

4. Tarr, *Search for the Ultimate Sink*, 385.

5. Brown, *Plutopia*, 59.

Chapter 1. The Secret Weapon

The chapter's epigraph is from Rhodes, *Making of the Atomic Bomb*, 500.

1. Bower, Rose, and Tighe, "Miracle with a Price"; Edward Mallinckrodt Jr. Papers.

2. "If Our Walls Could Talk."

3. The big four Allies in World War II were Great Britain, France, the Soviet Union, and the United States. The Soviet Union fought against Germany but did not declare war against Japan until 1945.

4. Bower, Rose, and Tighe, "Miracle with a Price."

5. "Mallinckrodt Group Inc." Today the town of Bremen is the Hyde Park neighborhood in St. Louis. Edward Mallinckrodt Papers.

6. "Nobel Prize."

7. Rhodes, *Making of the Atomic Bomb*, 363–64.

8. Rhodes, 364–65.

9. One important forerunner was German scientist Martin Klaproth, who in 1789 discovered the element uranium in oxide form in the mineral pitchblende. Then, in 1896, French physicist Henri Becquerel discovered that uranyl potassium sulfate, a uranium salt, emits energetic and penetrating radiation. Two years later, French scientists Marie and Pierre Curie discovered that thorium gives off "uranium rays," which Marie Curie renamed "radioactivity."

10. This is the process by which neutrons released in the splitting of an atomic nucleus produce an additional splitting in at least one further atomic nucleus. This nucleus in turn produces neutrons, and the process repeats.

11. Rhodes, "Explosive Discoveries and Bureaucratic Inertia," 17–18; U.S. Department of Energy, Manhattan Project.

12. "Einstein's Letter"; "Albert Einstein to F. D. Roosevelt."

13. Susan Williams, *Spies in the Congo*, xxiii–xxiv, 1–2.

14. Kevin O'Neill, "Building the Bomb," 53.

15. Kevin O'Neill, 54.

16. Kevin O'Neill, 54, 55.

17. Kevin O'Neill, 55.

18. Kelly, *Manhattan Project*, 69.

19. Susan Williams, *Spies in the Congo*, 5; "Uranium Mining."

20. Kelly, *Manhattan Project*, 69–70.

21. Susan Williams, *Spies in the Congo*, 4.

22. Gerber, *On the Home Front*, 23–24.

23. Rhodes, *Making of the Atomic Bomb*, 426.

24. Kelly, *Manhattan Project*, 69–70.

25. Kevin O'Neill, "Building the Bomb," 59; Schwartz, "Congressional Oversight of the Bomb," 485.

26. Susan Williams, *Spies in the Congo*, 10.

27. Bower, Rose, and Tighe, "Miracle with a Price"; Edward Mallinckrodt Papers; "Mallinckrodt Group, Inc."

28. Fermi, "Chicago Pile-1," 85.

29. Bower, Rose, and Tighe, "Miracle with a Price."

30. Susan Williams, *Spies in the Congo*, 2–3.

31. Susan Williams, 2–3; Vincent C. Jones, *Manhattan*, 8, 24.

32. Vincent C. Jones, 79–80.

33. Vincent C. Jones, 300.

34. *St. Louis Site Remediation Task Force Report*, 9; Swain, "Congo's Role in Creating."

35. Vincent C. Jones, *Manhattan*, 300–301.

36. Lanquette, "Enlisting Einsteins," 38; "Albert Einstein to F. D. Roosevelt," 43; "Uranium Mining"; Susan Williams, *Spies in the Congo*, 7.

37. Rhodes, *Making of the Atomic Bomb*, 355; Kelly, *Manhattan Project*, 155.

38. "Process Used in 1942." The process began by putting ether at 32 degrees Fahrenheit into an extractor. Next came the heating of a crude yellow-crystal form of uranium (uranyl nitrate) into liquid form and adding it at 176 degrees Fahrenheit. (Heat exchangers with ice water kept the mixture cool.) During a mixing process, the uranyl nitrate was extracted into ether. Most impurities were drained to the waste tank, and other impurities were washed from the ether solution. What followed was the mixing of distilled water with the solution of ether and uranyl nitrate. Because there were no longer any impurities, the uranyl nitrate was extracted from ether into the water. The solution of water and uranyl nitrate then went into the product tank to await further processing. Most of the ether stayed in the extraction.

39. U.S. Department of Energy, Manhattan Project; "History of the St. Louis Uranium Processing," 2–3; "Oak Ridge, TN."

40. "History of the St. Louis Uranium Processing," 2–3.

41. Bower, Rose, and Tighe, "Miracle with a Price."

42. Bower, Rose, and Tighe, "Miracle with a Price."

43. Bower, Rose, and Tighe, "Miracle with a Price."

44. Kelly, *Manhattan Project*, 317.

45. Rhodes, *Making of the Atomic Bomb*, 676; Levy, "Test That Changed the World."

46. Kelly, *Manhattan Project*, 317; "Very Sobering Event," 329.

47. Truman, "Battle of the Laboratories," 339–40; Susan Williams, *Spies in the Congo*, 242.

48. Kelly, *Manhattan Project*, 318.

49. Bower, Rose, and Tighe, "Miracle with a Price."

50. "World War II and St. Louis."

51. Tim O'Neill, "Look Back: St. Louis Factory" and "Look Back: How 35,000 St. Louis Workers."

52. Kelly, *Manhattan Project*, 318; Weller, "Atomic Bomb's Peculiar 'Disease,'" 356–69.

53. Broad, "Truth behind the News."

54. Hersey, *Hiroshima*.

55. Rhodes, *Making of the Atomic Bomb*, 734.

56. Kelly, *Manhattan Project*, 363; Hersey, "Hersey's *Hiroshima*,'" 377–81.

57. Stimson, "Decision to Use the Atomic Bomb," in *Manhattan Project*, ed. Kelly, 383–88; Stimson, "Decision to Use the Atomic Bomb," *Harper's Magazine*, February 1947, 97–107.

58. Walker, "Historiographical Essay," 311–34. Walker was also the author of *Prompt and Utter Destruction: Truman and the Use of Atomic Bombs against Japan.*

59. Walker, "Historiographical Essay," 312. "Traditional" scholars and their works included Allen and Polmar, *Code-Name Downfall*; Ferrell, *Harry S. Truman*; Robert P. Newman, *Truman and the Hiroshima Cult*; Maddox, *Weapons for Victory*; and Giangreco, *Hell to Pay*.

60. Walker, "Historiographical Essay," 312. "Revisionist" scholars and their works included Alperovitz, who wrote *Atomic Diplomacy* and *Decision to Use the Atomic Bomb*; Bird and Lifschultz, *Hiroshima's Shadow*; Lifton and Mitchell, *Hiroshima in America*; Skates, *Invasion of Japan*; and Wainstock, *Decision to Drop the Atomic Bomb*.

61. Walker, "Historiographical Essay," 319.

62. Walker, 324–34. "Middle-ground" scholars and their works included Bernstein, "Understanding the Atomic Bomb and the Japanese Surrender"; Frank, *Downfall*; and Zeiler, *Unconditional Defeat*. Scholars whose works were published after 2005 and were not categorized by Walker include Campbell and Radchenko, *Atom Bomb and the Origins of the Cold War*; Dower, *Cultures of War*; Gordin, *Five Days in August*; Malloy, *Atomic Tragedy*; Miscamble, *Most Controversial Decision*; and Rotter, *Hiroshima: The World's Bomb*.

63. Freshwater, "Nuclear Waste Creates Casualties."

64. Thorium is a weakly radioactive, metabolic chemical element. It will not itself split and release energy. Rather, when it is exposed to neutrons, it

will undergo a series of nuclear reactions until it eventually emerges as an isotope called U-233, which will be ready to split and release energy the next time it absorbs a neutron.

65. Bower, Rose, and Tighe, "Some Feared for the Health."

66. Bower, Rose, and Tighe, "Some Feared for the Health."

67. Bower, Rose, and Tighe, "Some Feared for the Health"; "Uranium and Depleted Uranium."

68. Bower, Rose, and Tighe, "Some Feared for the Health."

Chapter 2. The Deposit

The chapter's epigraph is from Koenig, "Cleanup Pledge by Bush."

1. *St Louis Site Remediation Task Force Report*, II-1.

2. Gilbert, "After 50 Years, Radioactive Waste."

3. *St. Louis Site Remediation Task Force Report*, II-1-3; Kay Drey Mallinckrodt Collection.

4. *St. Louis Site Remediation Task Force Report*, II-1-3.

5. Kevin O'Neill, "Building the Bomb," 61–62.

6. Kevin O'Neill, 13–19.

7. Wammack, "Atomic Governance," 89.

8. Makhijani, Schwartz, and Weida, "Nuclear Waste Management," 357.

9. Schein, *Organizational Cultures and Leadership*.

10. Makhijani, Schwartz, and Weida, "Nuclear Waste Management," 356–67.

11. Kay Drey Mallinckrodt Collection; Bower, Rose, and Tighe, "Building a Mountain."

12. *St. Louis Site Remediation Task Force Report*, 13–19.

13. "Tom Green Interview—Excerpts"; Bower, Rose, and Tighe, "Building a Mountain."

14. "Tom Green Interview—Excerpts"; "Coldwater Creek Facts."

15. "Radioactive Dirt on Highway."

16. Bower, Rose, and Tighe, "Building a Mountain."

17. Bower, Rose, and Tighe, "Building a Mountain."

18. Kay Drey Mallinckrodt Collection.

19. Middleton, "Drainage Map"; "Mississippi River Facts."

20. *St. Louis Site Remediation Task Force Report*, II-1-3.

21. Bower, Rose, and Tighe, "Building a Mountain."

22. Missouri Department of Natural Resources Water Protection Program, *Bacteria Total Maximum Daily Load*, 3.

23. *Record of Decision*, 2–16; *Coldwater Creek, Missouri*, 6; *Atomic Homefront* film website.

24. Allan M. Jones, *Environmental Biology*, 44.

25. Garraghan, *Saint Ferdinand de Florissant*, 12.

26. "FUSRAP: St. Louis Downtown Site."

27. Fentem, "Cleanup of Manhattan Project Site."

28. Swain, "Congo's Role in Creating"; "Making the Bomb."

29. *St. Louis Site Remediation Task Force Report*, II-1.

30. *Missouri State Gazetteer*, 263; *Journal of the Missouri Constitutional Convention*, 82. See also the *Journal of the House of Representatives*, 166; *Pitzman's New Atlas*; "Loyalty Oath"; "Judge James C. Edwards, 1883"; "Judge James C. Edwards Buried"; Tim O'Neill, "After the Great Divorce."

31. Corbett, "Draining the Metropolis," 117–18; Gordon, *Mapping Decline*, 22.

32. Primm, *Lion of the Valley*, 298–300; Wanko, "Great Divorce"; Tim O'Neill, "August 22, 1876"; Cooperman, "St. Louis' Great Divorce."

33. E. Terrence Jones, *Fragmented by Design*, 1–3.

34. Kalin, "History of Ferguson, Missouri," 67.

35. Hurley, "Busby's Stink Boat," 148.

36. Hurley, "Busby's Stink Boat," 147.

37. Hurley, "Busby's Stink Boat," 147–62.

38. Hurley, *Common Fields*, 2.

39. Hurley, "Busby's Stink Boat," 145 57.

40. Hurley, "Busby's Stink Boat," 156–59.

41. Hurley, "Busby's Stink Boat," 160–62.

42. Rust, *Aerial Crossroads of America*, 7–18, 43–44; Schroeder, "Environmental Setting," 23.

43. Rust, *Aerial Crossroads of America*, 15–16.

44. Rust, 17; "Leadership."

45. Rust, *Aerial Crossroads of America*, 43.

46. Rust, 113.

47. Rust, 110–11.

48. Rust, 59.

49. Rust, 77.

50. Rust, 77–78.

51. Rust, 111.

52. Primm, *Lion of the Valley*, 447; Tarr and Zimring, "Struggle for Smoke Control," 200.

53. Tarr and Zimring, "Struggle for Smoke Control," 215, 217–18, 220.

54. Tarr and Zimring, 220.

55. Koenig, "Cleanup Pledge by Bush"; "Radioactive Dirt on Highway."

56. Tarr, *Search for the Ultimate Sink*, 336.

57. Andrews, *Managing the Environment, Managing Ourselves*, 383–91.

58. Hurley, "Busby's Stink Boat," 147.

59. Alvarez, "West Lake Story"; Wohlforth, "What Lies beneath the Fernald Preserve."

60. "Edgar Sengier."

61. Kay Drey Mallinckrodt Collection.

62. Kay Drey Mallinckrodt Collection.

63. Kay Drey Mallinckrodt Collection.

Chapter 3. The Contamination Spreads

The chapter's epigraph is from Freshwater, "Nuclear Waste Creates Casualties."

1. "Mayor Arthur F. Bangert."

2. Primm, *Lion of the Valley*, 445.

3. "Mayor Arthur F. Bangert"; *St. Louis County Postwar Subdivisions Study*.

4. Although there is some dispute over whether the town's original name was St. Ferdinand or Florissant, it was previously called St. Ferdinand de Fleurissant, meaning "flourishing." Garraghan, *Saint Ferdinand de Florissant*, 11.

5. Florissant, Missouri, Document Center.

6. "Florissant," St. Louis Realtors; Tim O'Neill, "Look Back: St. Louis Suburbs."

7. Andrews, *Managing the Environment, Managing Ourselves*, 179–200.

8. Andrews, 186–88.

9. Andrews, 179, 182–84.

10. Andrews, 184.

11. Rome, *Bulldozer in the Countryside*, 39.

12. Rome, 39–40.

13. E. Terrence Jones, *Fragmented by Design*, 25–26.

14. Ammon, *Bulldozer*, 6, 97–98.

15. Ammon, 7.

16. "Early Prairies of St. Louis"; Schroeder, "Environmental Setting," 33–35; Krusekopf and Pratapas, "Soil Survey."

17. "Early Prairies of St. Louis."

18. Franzwa, *History of the Hazelwood School District*, 2.

19. Allan M. Jones, *Environmental Biology*, 39.

20. Franzwa, *History of the Hazelwood School District*, 6.

21. Franzwa, 96.

22. E. Terrence Jones, *Fragmented by Design*, 26.

23. Rome, *Bulldozer in the Countryside*, 13.

24. When Mayer-Raisher-Mayer envisioned a new subdivision east of Florissant's original street grid, Paddock Hills was an upscale residential neighborhood in Cincinnati. Built between 1919 and 1957, the neighborhood was named for Albert Paddack, an early pioneer and probate judge in Ohio. Interestingly, Herman B. Mayer (president of Mayer-Raisher-Mayer) and his son and partner Alfred Mayer both lived in Cincinnati in 1940. The third partner, Irvin Raisher, was Herman Mayer's son-in-law. "Paddock Hills—Cincinnati"; "Decennial Census Official Publications" (1940).

25. Primm, *Lion of the Valley*, 478.

26. Advertisement for Paddock Hills; "Display Homes"; "Paddock Hills, Florissant."

27. Rome, *Bulldozer in the Countryside*, 3.

28. Rome, 40–43.

29. Andrews, *Managing the Environment, Managing Ourselves*, 196–99; Rome, *Bulldozer in the Countryside*, 3–4.

30. *Coldwater Creek, Missouri*, 6.

31. Discover the City; "Village of Hazelwood."

32. Hazelwood was the plantation home of Richard Graham and Catherine Mullanphy Graham, who married in 1824. The property was passed down through their family. Its last resident was Hattie Frost Fordyce, a daughter of Confederate general Daniel Marsh Frost. In 1953 she gave the home to St. Louis University, and it was demolished for industrial use ten years later.

33. E. Terrence Jones, *Fragmented by Design*, 27.

34. E. Terrence Jones, 27.

35. E. Terrence Jones, 29–30.

36. Franzwa, *History of the Hazelwood School District*, 1, 98–112.

37. Emrich, "History of Ferguson-Florissant," 59–60.

38. "Classification of Municipalities."

39. Freshwater, "Nuclear Waste Creates Casualties."

40. Gordon, *Mapping Decline*, 71–98; Herbold, "It Was Never," 104–8.

41. Bower, Rose, and Tighe, "Building a Mountain."

42. *Record of Decision*, 2–11.

43. *Record of Decision*, 2–34.

44. Garraghan, *Saint Ferdinand de Florissant*, 11–12.

Chapter 4. Bureaucratic Blues

The chapter's epigraph is from Harrison, "Early Atomic Waste Lingers."

1. Walker, "Short History of Nuclear Regulation," 4.

2. Makhijani, Schwartz, and Weida, "Nuclear Waste Management," 357.

3. Bower, Rose, and Tighe, "Contamination."

4. *St. Louis Site Remediation Task Force Report*, II-4; Alvarez, "West Lake Story."

5. Alvarez, "West Lake Story"; *St. Louis Site Remediation Task Force Report*, 40–41; *Record of Decision*, 2–4.

6. Alvarez, "West Lake Story."

7. Kay Drey Mallinckrodt Collection; U.S. Department of Health and Human Services, *Evaluation of Community Exposures*, iv.

8. *St. Louis Site Remediation Task Force Report*, 40–41; *Record of Decision*, 2–4.

9. Gilbert, "After 50 Years, Radioactive Waste."

10. *St. Louis Site Remediation Task Force Report*, 42–44.

11. *St. Louis Site Remediation Task Force Report*, 43.

12. Merchant, *Major Problems*, 481.

13. Andrews, *Managing the Environment, Managing Ourselves*, 228–29.

14. *St. Louis Site Remediation Task Force Report*, 39.

15. *St. Louis Site Remediation Task Force Report*.

16. *Record of Decision*, 2–3 to 2–4; "FUSRAP: SLAPS Slide Show."

17. *St. Louis Site Remediation Task Force Report*, 43–44; Alvarez, "West Lake Story."

18. *St. Louis Site Remediation Task Force Report*, 43–44.

19. *St. Louis Site Remediation Task Force Report*, II-4; *Record of Decision*, 2–4.

20. *St. Louis Site Remediation Task Force Report*, II-4; Clay to Graham, September 13, 1835. Major Richard Graham was an Indian agent and former aide to William Henry Harrison in the War of 1812. His wife, Catherine, was the daughter of John Mullanphy, reputed to be Missouri's first millionaire. Richard Graham Papers; "Death of the Venerable"; Rice, "Fordyce Mansion."

21. *St. Louis Site Remediation Task Force Report*, 46.

22. Alvarez, "West Lake Story."

23. *St. Louis Site Remediation Task Force Report*, 45–47.

24. This is a retaining wall made of stacked stone-filled containers connected with wire.

25. *St. Louis Site Remediation Task Force Report*, 46.

26. *Coldwater Creek, Missouri*, 1–2.

27. *St. Louis Site Remediation Task Force Report*, 25.

28. *St. Louis Site Remediation Task Force Report*, 47–48.

29. *St. Louis Site Remediation Task Force Report*, 48.

30. *St. Louis Site Remediation Task Force Report*, "Mission Statement."

31. *St. Louis Site Remediation Task Force Report*, ES-1.

32. *St. Louis Site Remediation Task Force Report*, ES-2.

33. *St. Louis Site Remediation Task Force Report*, III-4.

34. Radioactivity in the soil is usually measured in picocuries per gram. The standard was "thorium/radium concentrations not to exceed 5 picocuries per gram (5 pCi/g) averaged over the first 15 cm of soil and 15 picocuries per gram (15 pCi/g) averaged over 15 cm thick layers of soil more than 15 cm below the surface." Agency for Toxic Substances and Disease Registry (ATSDR) Public Health Statements.

35. *St. Louis Site Remediation Task Force Report*, II-6.

36. *St. Louis Site Remediation Task Force Report*, II-16.

37. *St. Louis Site Remediation Task Force Report*, III-7.

38. *St. Louis Site Remediation Task Force Report*, III-32-33.

39. *St. Louis Site Remediation Task Force Report*, III-28.

40. U.S. Army Corps of Engineers, St. Louis District FUSRAP.

41. *Record of Decision*.

42. *Record of Decision*.

43. Rehg, "Florissant City Council."

44. *St. Louis Site Remediation Task Force Report*, 47.

45. Harrison, "Early Atomic Waste Lingers"; Schneider, "Mountain of Nuclear Waste."

46. Harrison, "Early Atomic Waste Lingers."

47. Bower, Rose, and Tighe, "Solution: Redrawing Maps."

48. "Chance Sparked Crusade."

49. "Chance Sparked Crusade."

50. "Chance Sparked Crusade."

51. "Chance Sparked Crusade."

52. Kay Drey Mallinckrodt Collection.

53. "Transfer of Radioactive Dirt."

54. "Radioactive Dirt on Highway."

55. Hartmann, "Publisher's Column."

56. Tighe, "Carolyn Bower Remembers Lou Rose."

57. Walker, *Permissible Dose*, 8–11.

58. *St. Louis Site Remediation Task Force Report*, II-6.

59. Walker, *Permissible Dose*, 10–11.

60. Walker, *Permissible Dose*, 16.

61. Walker, *Permissible Dose*, 16–17; Makhijani and Schwartz, "Victims of the Bomb," 427–28.

62. Walker, *Permissible Dose*, 10–11.

63. Gilbert, "Standards to Remove Hazardous Waste."

64. "St. Louis Airport Site."

65. Clusen to Young, January 16, 1979. President Jimmy Carter did not sign the U.S. Department of Energy Organization Act until August 4, 1977, which created the DOE as a cabinet-level department. Carter's election to the presidency occurred in November 1976, so the radiological site survey that Clusen discussed would have occurred in the final months of the Ford administration. Nevertheless, her January 1979 letter to Young confirms DOE ownership of the report, which was then "in a final review stage prior to its publication."

66. Clusen to Young, January 16, 1979.

67. Clusen to Young, January 16, 1979.

68. "Radionuclide Basics."

69. Gilbert, "After 50 Years, Radioactive Waste."

70. Gilbert.

71. Vierzba to Whitman, August 20, 1985.

72. Vierzba to Whitman, August 20, 1985.

73. Susan Williams, *Spies in the Congo*, 3–5.

74. Vierzba to Whitman, August 20, 1985.

Chapter 5. The Advocates

The chapter's epigraph is from Ratcliffe, *Oxford Essential Quotations*.

1. Wright, interview with author, February 5, 2020.

2. Freshwater, "Nuclear Waste Creates Casualties."

3. Freshwater.

4. Freshwater.

5. Hartmann, "We Wrote about Poisons."

6. Freshwater, "Nuclear Waste Creates Casualties."

7. Wright, interview with author, February 5, 2020.

8. Alvarez, "West Lake Story."

9. Freshwater, "Nuclear Waste Creates Casualties."

10. Siouxland News at Sunrise.

11. U.S. Environmental Protection Agency, "National Environmental Education Act"; "An Act to Promote Environmental Education."

12. "What Is Environmental Education?"; National Service Center for Environmental Publications, *Fact Sheet.*

13. Armstrong, "Bloom's Taxonomy"; Bloom, *Taxonomy of Educational Objectives*; North American Association for Environmental Education.

14. Freshwater, "Nuclear Waste Creates Casualties."

15. Freshwater.

16. Yun et al., *Analysis of Cancer Incidence Data*; Lupkin, "Dispute over Missouri Cancer Cluster."

17. Schanzenbach and Wright, "Commentary."

18. Schanzenbach, "Cancer Study."

19. Freshwater, "Nuclear Waste Creates Casualties."

20. Freshwater.

21. Freshwater; Emshwiller, "Nuclear Waste Taints"; "7 More Nuclear Waste 'Hot Spots.'"

22. Emshwiller, "Nuclear Waste Taints."

23. Emshwiller.

24. For example, Coldwater Creek is depicted as a single line in the May 20, 2013, issue of the *St. Louis Post-Dispatch.* A more accurate representation appears in Leland D. Hauth and Donald W. Spencer, *Floods in Coldwater Creek.* Published in 1971, it shows several tributaries feeding into the waterway.

25. U.S. Department of Health and Human Services, *Evaluation of Community Exposures*, iii, 21.

26. U.S. Department of Health and Human Services, *Evaluation of Community Exposures*, iv.

27. U.S. Department of Health and Human Services, *Evaluation of Community Exposures*.

28. Hartmann, "Poisoned Children of Coldwater Creek."

29. U.S. Department of Health and Human Services, *Evaluation of Community Exposures*, iv.

30. Khan, "Health Advisory"; Khan, "Legacy of Environmental Health Concerns."

31. Hartmann, "We Wrote about Poisons."

32. Hartmann, "We Wrote about Poisons."

33. *Atomic Homefront* film website; First Secret City.

34. Fentem, "Residents Say Coldwater Creek Report."

35. Fentem, "Residents Say Coldwater Creek Report"; Freshwater, "Nuclear Waste Creates Casualties"; Steingraber, *Living Downstream*, 73–79.

36. Freshwater, "Nuclear Waste Creates Casualties."

37. "Coldwater Creek Facts."

38. Zaretsky, *Radiation Nation*, 104.

39. Zaretsky, 126.

40. Newman, *Love Canal*, 132–33.

41. Bell, "Bread and Roses."

42. Just Moms STL.

43. "West Lake Landfill"; Price, "Radioactive Waste."

44. "West Lake Landfill."

45. Price, "Radioactive Waste."

46. Gray, "EPA Wants to Do."

47. Raasch, "Plans to Shift."

48. "Just Moms STL"; Johnson, "Fallout."

49. Kathleen Logan Smith, "Coalition Report."

50. Danielson, "Families Sue"; Davis, "Tempers Flare at Meeting"; Chen, "After 'Atomic Homefront' Release."

51. *Coldwater Creek Lawsuit.* For more information about the lawsuits, see the following articles by Blythe Bernhard in the *St. Louis Post-Dispatch*: "Lawsuit Links Illnesses"; "North St. Louis County Group"; "Judge Throws Out." See also Kirn, "Another Lawsuit Accuses Mallinckrodt"; Bouscaren, "Former McDonnell Douglas Workers"; Schuessler, "Missouri Home Contaminated"; "St. Louis Area Residents Sue"; and Thomas, "St. Louis Man's Alleged Radiation."

52. Patrick, "Bridgeton Landfill Agrees."

53. Salter, "West Lake Landfill Cleanup Slows."

54. "St. Louis Site Fact Sheet."

55. Bogan, "It's Just Been Ridiculous."

56. Bogan.

57. Steingraber, *Living Downstream*, 15.

58. Masco, *Nuclear Borderlands*, vii.

Chapter 6. An Environmental Justice Watershed

1. Missouri Department of Natural Resources Water Protection Program, *Bacteria Total Maximum Daily Load*, 7; *Studies of Aquatic Life*; "Office of Environmental Justice."

2. Bullard, "Solid Waste Sites," 273–88.

3. Hurley, "Floods, Rats, and Toxic Waste," 253.

4. Been, "Locally Undesirable Land Uses," 1383–1422.

5. Lambert and Boerner, "Environmental Inequity."

6. Lambert and Boerner, 202–6.

7. Lambert and Boerner, 202.

8. Hurley, "Floods, Rats, and Toxic Waste," 260.

9. Sarathy, Hamilton, and Brodie, *Inevitably Toxic*, 13.

10. Foley, *History of Missouri*, 1:46–47, 175–76; McCandless, *History of Missouri*, 2:37; Garraghan, *Saint Ferdinand de Florissant*, 60–74; Richard Graham Papers; Kruse, *Old Jamestown across the Ages*, 112–13; Florissant Valley Historical Society at Taille de Noyer.

11. *St. Louis Site Remediation Task Force Report*, II-4; Clay to Graham, September 13, 1835; "Death of the Venerable"; Rice, "Fordyce Mansion."

12. Theising, *In the Walnut Grove*, 65. Note: "Old St. Ferdinand Township"

refers to the land bounded on the north by the Missouri River, on the east by the Mississippi River and St. Louis City, on the south by Central Township, and on the west by Bonhomme Township and the Missouri River. Although Old St. Ferdinand was one of five townships in St. Louis County when it separated from the City, many more currently exist. It should not be confused with the St. Ferdinand Township of today, which is a different geographic location and not relevant to this book.

13. Wright, *St. Louis*, 115; Irene Sanford Smith, *Ferguson*, 33–35.

14. Wright, *St. Louis*, 89–101, 117–19.

15. Wright, *St. Louis*, 67.

16. Wright, *St. Louis*, 89, 90, 117, 118; Irene Sanford Smith, *Ferguson*, 35.

17. Wright, *St. Louis*, 102.

18. Wright, *Kinloch*, 9–22.

19. Hamilton and Webb, "Historic Buildings Survey," 5; Wright, *Kinloch*, 29–33.

20. Gordon, *Mapping Decline*, 146.

21. "Decennial Census Official Publications" (1960); "Decennial Census Official Publications" (1970).

22. Wright, *Kinloch*, 51.

23. Tuft, "Airport Seeks All of Kinloch."

24. "Decennial Census Official Publications" (1990); "Decennial Census Official Publications" (2000).

25. Benchaabane, "Looking for Help in Kinloch."

26. Primm, *Lion of the Valley*, 478.

27. Beine, "Population of St. Louis City."

28. Primm, *Lion of the Valley*, 478–79, 485.

29. Gordon, *Mapping Decline*, 83–84.

30. Gordon.

31. Gordon, 86.

32. Gordon, 146.

33. Gordon, 22.

34. Cooperman, "Story of Segregation."

35. Gordon, *Mapping Decline*, 102; "Civil Rights Act of 1866."

36. "Decennial Census Official Publications" (1990).

37. Hurley, "Floods, Rats, and Toxic Wastes," 258–61.

38. Barker, "Residents of Coldwater Creek"; *St. Louis Site Remediation Task Force Report*, III-4; "FUSRAP Questions: St. Louis District" (appendix C, this volume), question 30.

39. "7 Things You Should Know."

40. Franzwa, *History of the Hazelwood School District*, 96; *Coldwater Creek, Missouri*; *Record of Decision*, 2–10.

41. United States v. State of Missouri, 388 F. Supp. 1058 (E.D. Mo 1975).

42. "Florissant, Missouri Population History"; "Hazelwood, Missouri Population History"; "Berkeley, Missouri Population History."

43. U.S. Department of Health and Human Services, *Evaluation of Community Exposures.*

44. World Population Review; Suburban Stats; Fontinelle, "Average Home Price by State." Precise comparisons cannot be made in the categories of home price and household income. The *median* is the middle value in a distribution of numbers. It differs from the arithmetic *mean*, or average of the numbers.

45. Gay, "White Flight and White Power."

46. Barker, "Residents of Coldwater Creek."

47. See FUSRAP question 33 in appendix C.

Chapter 7. A Part of the Whole

The chapter's epigraph is from Hartmann, "Poisoned Children of Coldwater Creek."

1. Bernhard, "Judge Throws Out."

2. Freshwater, "Nuclear Waste Creates Casualties."

3. Bernhard, "Nurse Who Questioned."

4. Kevin O'Neill, "Building the Bomb," 33.

5. Kevin O'Neill, 36.

6. Kevin O'Neill, 36.

7. Sarathy, Hamilton, and Brodie, *Inevitably Toxic*, 7.

8. Terry Tempest Williams, *Refuge.*

9. Mason, "Fallout."

10. Iversen, *Full Body Burden.*

11. Pritikin, *Hanford Plaintiffs.*

12. Satterfield and Levin, "From Cold War Complex," 165.

13. Iversen, *Full Body Burden*, 231.

14. Gerber, *On the Home Front*, 1–2.

15. Brown, *Plutopia*, 15.

16. Gerber, *On the Home Front*, 25.

17. Brown, *Plutopia*, 16–18.

18. Freeman, *Longing for the Bomb*, 18.

19. Brown, *Plutopia*, 24–25.

20. Brown, 37.

21. Brown, 71.

22. For more information about the Native American experience at Hanford, see Liebow, "Hanford, Tribal Risks," 145–61.

23. Brown, *Plutopia*, 37–41.

24. Brown, 71.

25. Gerber, *On the Home Front*, 33.

26. Brown, *Plutopia*, 59.

27. Brown, 59.

28. Brown, 59.

29. Pritikin, *Hanford Plaintiffs*, 6–7; Zhang, "Secret 1949 Radiation Experiment." Green Run was a secret Air Force experiment. Normally irradiated fuel

is cooled for up to 101 days before it is processed, so short-lived radioactive elements like iodine can decay. In the Green Run, the fuel was cooled just six-
teen days. Officials told Washington's *Spokane Chronicle* that the "green" order
came from the military, who assumed Soviets were rushing to produce nuclear bombs. If they were short cooling the fuel, the radioactive result might
be spotted some distance away. So the U.S. Air Force wanted to fly planes beyond a radioactive plume to test out their instruments.

30. Pritikin, 10. (For a sense of scale, the Three Mile Island power accident
released an estimated fifteen to twenty-four curies of iodine-131, and the Cher-
nobyl accident released thirty-five to forty-nine million curries.)

31. Pritikin, 283–84.

32. Mason, "Fallout," 34.

33. Mason, 31.

34. "Historical Events"; Oncoi, "National Cancer Database"; "State Cancer
Registries."

35. U.S. Department of Health and Human Services, *Evaluation of Community Exposures*, 6.

36. Freshwater, "Nuclear Waste Creates Casualties."

BIBLIOGRAPHY

"An Act to Promote Environmental Education, and Other Purposes." Public Law 101-619. November 16, 1990.

Advertisement for Paddock Hills. *St. Louis (Mo.) Post-Dispatch*, May 20, 1956.

Agency for Toxic Substances and Disease Registry (ATSDR) Public Health Statements. https://wwwn.cdc.gov/TSP/PHS/PHSLanding.aspx.

"Albert Einstein to F. D. Roosevelt." In *The Manhattan Project: The Birth of the Atomic Bomb in the Words of Its Creators, Eyewitnesses, and Historians*, edited by Cynthia C. Kelly, 42–44. New York: Black Dog and Levanthal, 2009.

Allen, Thomas B., and Norman Polmar. *Code-Name Downfall: The Secret Plan to Invade Japan—and Why Truman Dropped the Bomb*. New York: Simon and Schuster, 1995.

Alperovitz, Gar. *Atomic Diplomacy: Hiroshima and Potsdam*. New York: Simon and Schuster, 1965.

———. *The Decision to Use the Atomic Bomb and the Architecture of an American Myth*. New York: Knopf, 1995.

Alvarez, Robert. "West Lake Story: An Underground Fire, Radioactive Waste, and Governmental Failure." *Bulletin of the Atomic Scientists*, February 11, 2016. https://thebulletin.org/2016/02/west-lake-story.

Ammon, Francesca Russello. *Bulldozer: Demolition and Clearance of the Postwar Landscape*. New Haven, Conn.: Yale University Press, 2016.

Andrews, Richard N. L. *Managing the Environment, Managing Ourselves: A History of American Environmental Policy*. New Haven, Conn.: Yale University Press, 1999.

Armstrong, Patricia. "Bloom's Taxonomy." Vanderbilt University Center for Teaching, 2010. https://cft.vanderbilt.edu/guides-sub-pages/blooms-taxonomy.

"Arthur Bangert, Mayor of Florissant, 1938–1950." The Historical Marker Database, accessed March 20, 2022. https://www.hmdb.org/m.asp?m=149393.

Atomic Homefront film website, 2017. https://www.atomichomefront.film.

Barker, Jacob. "Residents of Coldwater Creek Not Surprised at Nearby Contamination." *St. Louis (Mo.) Post-Dispatch*, August 20, 2015.

Been, Vicki. "Locally Undesirable Land Uses in Minority Neighborhoods: Disproportionate Siting or Market Dynamics?" *Yale Law Journal* 6 (April 1994): 1383–422.

Beine, Joe. "Population of St. Louis City & County, and Missouri, 1820–2010." Genealogy Articles, Tips and Research Guides, accessed March 19, 2022. https://www.genealogybranches.com/stlouispopulation.html.

Bell, Karen. "Bread and Roses: A Gender Perspective on Environmental Justice and Public Health." *International Journal of Environmental Research and Public Health* 13, no. 10 (2016). https://www.mdpi.com/1660-4601 /13/10/1005.

Benchaabane, Nassim. "Looking for Help in Kinloch." *St. Louis (Mo.) Post-Dispatch*, June 8, 2021.

"Berkeley, Missouri Population History," Biggest US Cities, updated January 25, 2022. https://www.biggestuscities.com/demographics/mo/berkeley -city.

Bernhard, Blythe. "Judge Throws Out Most Coldwater Creek Cancer Claims." *St. Louis (Mo.) Post-Dispatch*, March 29, 2013.

———. "Lawsuit Links Illnesses to North St. Louis County Creek." *St. Louis (Mo.) Post-Dispatch*, February 29, 2012.

———. "North St. Louis County Group Files 2nd Suit Alleging Nuclear Waste Caused Illnesses." *St. Louis (Mo.) Post-Dispatch*, April 11, 2012.

———. "Nurse Who Questioned Rare Cancer Link to Coldwater Creek Dies." *St. Louis (Mo.) Post-Dispatch*, August 7, 2014.

Bernstein, Barton. "Understanding the Atomic Bomb and the Japanese Surrender: Missed Opportunities, Little-Known Near Disasters, and Modern Memory." *Diplomatic History* 19, no. 2 (1995): 227–73.

Bilimora, Karl Y., Andrew K. Stewart, David P. Winchester, and Clifford Y. Ko. "The National Cancer Database: A Powerful Initiative to Improve Cancer Care in the United States." *Annals of Surgical Oncology* 15, no. 3 (March 2008): 683–90.

Bird, Kai, and Lawrence Lifschultz, eds. *Hiroshima's Shadow: Writings on the Denial of History and the Smithsonian Controversy*. Stony Creek, Conn.: Pamphleteer, 1998.

Bloom, Benjamin S., ed. *Taxonomy of Educational Objectives: Handbook 1; Cognitive Domain*. Boston: Addison-Wesley, 1956.

Bogan, Jesse. "It's Just Been Ridiculous." *St. Louis (Mo.) Post-Dispatch*, December 19, 2021.

Bouscaren, Durrie. "Former McDonnell Douglas Workers, Residents File Suit over Radiation Exposure." St. Louis Public Radio, May 20, 2016. https:// news.stlpublicradio.org/health-science-environment/2016-05-20/former -mcdonnell-douglas-workers-residents-file-suit-over-radiation-exposure.

Bower, Carolyn, Louis J. Rose, and Theresa Tighe. "Building a Mountain of Radioactive Waste." *St. Louis (Mo.) Post-Dispatch*, February 14, 1989.

———. "Contamination: How Weldon Springs Went from Model to Mess." *St. Louis (Mo.) Post-Dispatch*, February 12, 1989.

———. "A Miracle with a Price." *St. Louis (Mo.) Post-Dispatch*, February 12, 1989.

———. "Solution: Redrawing Maps." *St. Louis (Mo.) Post-Dispatch*, February 14, 1989.

———. "Some Feared for the Health of the Handlers." *St. Louis (Mo.) Post-Dispatch*, February 13, 1989.

Broad, William J. "The Truth behind the News." *New York Times*, August 10, 2021.

Brown, Kate. *Plutopia: Nuclear Families, Atomic Cities, and the Great Soviet and American Plutonium Disasters*. New York: Oxford University Press, 2013.

Bullard, Robert D. "Solid Waste Sites and the Black Houston Community." *Sociological Inquiry* 53 (April 1983): 273–88.

Campbell, Craig, and Sergey Radchenko. *The Atom Bomb and the Origins of the Cold War*. New Haven, Conn.: Yale University Press, 2008.

"Chance Sparked Crusade to Clean Up Waste Here." *St. Louis (Mo.) Post-Dispatch*, February 14, 1989.

Chen, Eli. "After 'Atomic Homefront' Release, Frustrated Residents Fill Army Corps Coldwater Creek Meeting." St. Louis Public Radio, February 23, 2018. https://news.stlpublicradio.org/health-science-environment/2018-02-23/after-atomic-homefront-release-frustrated-residents-fill-army-corps-coldwater-creek-meeting.

"Civil Rights Act of 1866." Owl Eyes website. https://www.owleyes.org/text/civil-rights-act-of-1866/read/text-of-the-act.

"Classification of Municipalities." Missouri Secretary of State–Missouri Roster, updated 2021–22. https://s1.sos.mo.gov/CMSImages/Publications/municipalities05.pdf.

Clay, Henry to Richard Graham, September 13, 1835. Henry Clay Letters. Mercantile Library Special Collections, University of Missouri–St. Louis.

Clusen, Ruth C. to Robert A. Young, January 16, 1979, Boxes 10 (102716) and 11 (102717). Kay Drey Mallinckrodt Collection. State Historical Society of Missouri, Columbia.

"Coldwater Creek Facts: About Us." Coldwater Creek Facts. www.coldwater-creekfacts.com/about-us.

Coldwater Creek Lawsuit—in Plain English (blog associated with "Coldwater Creek—Just the Facts Please" Facebook group), May 7, 2013. coldwater creekfacts.wordpress.com.

"Coldwater Creek Map." *St. Louis (Mo.) Post-Dispatch*, May 20, 2013.

Coldwater Creek, Missouri: Feasibility Report and Environmental Impact Statement. St. Louis, Mo.: U.S. Army Corps of Engineers, October 1986.

Cooperman, Jeanette. "St. Louis' Great Divorce: A Complete History of the City and County Separation and Attempts to Get Back Together." *St. Louis Magazine*, March 8, 2019.

———. "The Story of Segregation in St. Louis." *St. Louis Magazine*, October 17, 2014.

Corbett, Katharine. "Draining the Metropolis: The Politics of Sewers in Nine-
    teenth Century St. Louis." In *Common Fields: An Environmental History
    of St. Louis*, edited by Andrew Hurley, 107–25. St. Louis: Missouri Histori-
    cal Society Press, 1997.
Danielson, J. Ryne. "Families Sue over St. Louis' Radioactive Coldwater
    Creek." Patch.com, August 7, 2018. https://patch.com/missouri/stlouis
    /families-sue-over-st-louis-radioactive-coldwater-creek.
Davis, Chris. "Tempers Flare at Meeting over Coldwater Creek." KSDK.com,
    February 23, 2018. https://www.ksdk.com/article/news/local/tempers
    -flare-at-meeting-over-coldwater-creek/63-522355524.
"Death of the Venerable Richard Graham." *National Intelligencer*, August 8,
    1857.
"Decennial Census Official Publications" (1940). United States Census Bu-
    reau. https://www.census.gov/programs-surveys/decennial-census/decade
    /decennial-publications.1940.html.
"Decennial Census Official Publications" (1960). United States Census Bu-
    reau. https://www.census.gov/programs-surveys/decennial-census/decade
    /decennial-publications.1960.html.
"Decennial Census Official Publications" (1970). United States Census Bu-
    reau. https://www.census.gov/programs-surveys/decennial-census/decade
    /decennial-publications.1970.html.
"Decennial Census Official Publications" (1990). United States Census Bu-
    reau. https://www.census.gov/programs-surveys/decennial-census/decade
    /decennial-publications.1990.html.
"Decennial Census Official Publications" (2000). United States Census Bu-
    reau. https://www.census.gov/programs-surveys/decennial-census/decade
    /decennial-publications.2000.html.
"The Decision to Use the Atomic Bomb." In *The Manhattan Project: The
    Birth of the Atomic Bomb in the Words of Its Creators, Eyewitnesses, and
    Historians*, edited by Cynthia C. Kelly, 383–88. New York: Black Dog and
    Levanthal, 2009.
Discover Hazelwood, accessed July 28, 2020. www.hazelwoodmo.org/164
    /Discover-Hazelwood.
"Display Homes." *St. Louis (Mo.) Post-Dispatch*, May 20, 1956.
Dower, John W. *Cultures of War: Pearl Harbor / Hiroshima / 9-11 / Iraq*. New
    York: W. W. Norton, 2010.
"Early Prairies of St. Louis." Wild Ones St. Louis Chapter: Healing the Earth
    One Yard at a Time, December 8, 2011. https://stlwildones.org/early
    -prairies-of-st-louis/.
"Edgar Sengier." Atomic Heritage Foundation, accessed March 18, 2022.
    https://www.atomicheritage.org/profile/edgar-sengier.
"Einstein's Letter." Franklin D. Roosevelt Presidential Library and Museum,
    accessed March 18, 2022. http://www.fdrlibrary.marist.edu/archives/pdfs
    /docsworldwar.pdf.
Emrich, William. "A History of Ferguson-Florissant School District." In *His-

tory of Ferguson, compiled by Ferguson-Florissant School District, 227–314. Ferguson, Mo.: Ferguson-Florissant School District, 1975.

Emshwiller, John R. "Nuclear Waste Taints St. Louis Suburb." *Wall Street Journal*, August 23, 2015.

"Explosive Discoveries and Bureaucratic Inertia." In *The Manhattan Project: The Birth of the Atomic Bomb in the Words of Its Creators, Eyewitnesses, and Historians*, edited by Cynthia C. Kelly, 17–18. New York: Black Dog and Leventhal, 2009.

Fentem, Sarah. "Cleanup of Manhattan Project Site in Downtown St. Louis Nears Completion." St. Louis Public Radio, March 1, 2019. https://news .stlpublicradio.org/health-science-environment/2019-03-01/cleanup-of -manhattan-project-site-in-downtown-st-louis-nears-completion.

——. "Residents Say Coldwater Creek Report Lacks Answers to Cancer Questions." St. Louis Public Radio, June 26, 2018. https://news .stlpublicradio.org./health-science environment/2018-06-26 /residents-say-coldwater-creek-report-lacks-answers-to-cancer-questions.

Fermi, Enrico. "The Chicago Pile-1." In *The Manhattan Project: The Birth of the Atomic Bomb in the Words of Its Creators, Eyewitnesses, and Historians*, edited by Cynthia C. Kelly, 82–85. New York: Black Dog and Leventhal, 2009.

Ferrell, Robert H. *Harry S. Truman: A Life*. Columbia: University of Missouri Press, 1994.

The First Secret City, 2015. https://first-secret-city.com.

"Florissant, Missouri Population History." Biggest US Cities, accessed July 28, 2020. https://www.biggestuscities.com/city/florissant-missouri.

City of Florissant, Missouri, accessed March 23, 2022. https://florissantmo .com.

Florissant Valley Historical Society at Taille de Noyer, accessed March 23, 2022. https://florissantvalleyhs.com/.

Foley, William E. *A History of Missouri*. Vol. 1, *1673–1820*. Columbia: University of Missouri Press, 1971.

Fontinelle, Amy. "Average Home Price by State." The Ascent, 2020. https:// www.fool.com/the-ascent/research/average-house-price-state/.

Frank, Richard B. *Downfall: The End of the Imperial Japanese Empire*. New York: Penguin Books, 1999.

Franzwa, Gregory M. *History of the Hazelwood School District*. Hazelwood, Mo.: Board of Education of the Hazelwood School District, 1977.

Freeman, Lindsey A. *Longing for the Bomb: Oak Ridge and Atomic Nostalgia*. Chapel Hill: University of North Carolina Press, 2015.

Freshwater, Lori. "Nuclear Waste Creates Casualties of War in Missouri." *Earth Island Journal*, March 18, 2016. https://www.earthisland.org /journal/index.php/magazine/entry/casualties_of_war/.

"FUSRAP: SLAPS Slide Show." U.S. Army Corps of Engineers, St. Louis District Website, accessed March 23, 2022. https://www.mvs.usace.army.mil /Missions/FUSRAP/SLAPS/.

"FUSRAP: St. Louis Downtown Site." U.S. Army Corps of Engineers, St. Louis District Website, accessed March 23, 2022. https://www.mvs.usace .army.mil/Missions/FUSRAP/SLDS/.

Garraghan, Gilbert J., S.J. *Saint Ferdinand de Florissant: The Story of an Ancient Parish*. Chicago: Loyola University Press, 1923.

Gay, Malcolm. "White Flight and White Power in St. Louis." *Time*, August 13, 2014.

Gerber, Michele Stenehjem. *On the Home Front: The Cold War Legacy of the Hanford Nuclear Site*. Lincoln: University of Nebraska Press, 2007.

Giangreco, D. M. *Hell to Pay: Invasion of Japan, 1945–1947*. Annapolis, Md.: Naval Institute Press, 2009.

Gilbert, Virginia Baldwin. "After 50 Years, Radioactive Waste at the Airport Is Finally Being Removed." *St. Louis (Mo.) Post-Dispatch*, April 30, 2000.

———. "Standards to Remove Hazardous Waste Weren't Developed until 1980." *St. Louis (Mo.) Post-Dispatch*, April 30, 2000.

Gordin, Michael D. *Five Days in August: How World War II Became a Nuclear War*. Princeton, N.J.: Princeton University Press, 2007.

Gordon, Colin. *Mapping Decline: St. Louis and the Fate of the American City*. Philadelphia: University of Pennsylvania Press, 2008.

Graham, Richard. Papers, 1795–1896. Missouri Historical Society Archives, St. Louis.

Gray, Bryce. "EPA Wants to Do Partial Excavation of Contaminants at Radioactive West Lake Landfill Superfund Site." *St. Louis (Mo.) Post-Dispatch*, February 1, 2018.

Hamilton, Esley, and Mary Webb. "Historic Buildings Survey Schools Built before 1941 in Saint Louis County." Missouri State Parks website, accessed March 23, 2002. mostateparks.com/sites/mostateparks/files/STLC %20Schools%20Report.pdf.

Harrison, Eric. "Early Atomic Waste Lingers in St. Louis Suburbs." *Los Angeles Times*, February 23, 1991.

Hartmann, Ray. "The Poisoned Children of Coldwater Creek Finally Get a Break." *St. Louis Magazine*, August 3, 2018. https://www.stlmag.com /news/think-again/the-poisoned-children-of-coldwater-creek-finally -get-a-break/.

———. "Publisher's Column." *Riverfront Times* (St. Louis, Mo.), August 4, 1982.

———. "We Wrote about Poisons in Coldwater Creek 37 Years Ago. Guess What the Feds Just Confirmed?" *Riverfront Times* (St. Louis, Mo.), May 15, 2019. https://www.riverfronttimes.com/stlouis/we-wrote-about-poisons-in-coldwater-creek-37-years-ago-guess-what-the-feds-just-confirmed.

Hauth, Leland D., and Donald W. Spencer. *Floods in Coldwater Creek, Watkins Creek, and River des Peres Basin, St. Louis County, Missouri*. St. Louis: U.S. Department of the Interior Geological Survey, 1971. Prepared in cooperation with Metropolitan Sewer District, St. Louis, Mo. https:// pubs.usgs.gov/of/1971/0146/report.pdf.

"Hazelwood, Missouri Population History." Biggest US Cities, accessed March 19, 2022. https://www.biggestuscities.com/city/hazelwood-missouri.

Herbold, Hilary. "It Was Never a Level Playing Field: Blacks and the GI Bill." *Journal of Blacks in Higher Education* (Winter 1994/1995): 104–8.

Hersey, John. "Hersey's *Hiroshima*." In *The Manhattan Project: The Birth of the Atomic Bomb in the Words of Its Creators, Eyewitnesses, and Historians*, edited by Cynthia C. Kelly, 377–81. New York: Black Dog and Leventhal, 2009.

——— . *Hiroshima*. New York: Alfred A. Knopf, 1946.

"Historical Events." National Cancer Institute: SEER Training Modules, accessed March 19, 2022. https://training.seer.cancer.gov/registration/registry/history/dates.html.

"The History of the St. Louis Uranium Processing Plant Radioactive Waste Sites." St. Louis Site Remediation Task Force. In *St. Louis Site Remediation Task Force Report*, C-1–C-32. St. Louis, Mo.. U.S. Army Corps of Engineers, St. Louis District, 1996.

Hurley, Andrew, ed. "Busby's Stink Boat and the Regulation of Nuisance Trades, 1865–1918." In *Common Fields: An Environmental History of St. Louis*, edited by Andrew Hurley, 145–62. St. Louis: Missouri Historical Society Press, 1997.

——— . *Common Fields: An Environmental History of St. Louis*. St. Louis: Missouri Historical Society Press, 1997.

——— . "Floods, Rats, and Toxic Waste: Allocating Environmental Hazards since World War II." In *Common Fields: An Environmental History of St. Louis*, edited by Andrew Hurley, 242–61. St. Louis: Missouri Historical Society Press, 1997.

"If Our Walls Could Talk." Lawrence Group, April 24, 2017. http://www.thelawrencegroup.com/if-our-walls-could-talk/.

Iversen, Kristen. *Full Body Burden: Growing Up in the Nuclear Shadow of Rocky Flats*. New York: Broadway Paperbacks, 2013.

Johnson, Lacy M. "The Fallout." *Guernica*, July 10, 2017. https://www.guernicamag.com/the-fallout/.

Jones, Allan M. *Environmental Biology*. London: Routledge, 1996.

Jones, E. Terrence. *Fragmented by Design: Why St. Louis Has So Many Neighborhoods*. St. Louis: Palmerston and Reed, 2000.

Jones, Vincent C. *Manhattan: The Army and the Atomic Bomb*. Washington, D.C.: Center of Military History, U.S. Army, 1985.

Journal of the House of Representatives of the State of Missouri. Jefferson City, Mo.: Cheeney, 1860.

Journal of the Missouri Constitutional Convention of 1875. Vol. 1. Columbia: State Historical Society of Missouri, 1920.

"Judge James C. Edwards Buried at Fee Fee Cemetery." *St. Louis (Mo.) Post-Dispatch*, December 15, 1883.

"Judge James C. Edwards, 1883." *St. Louis (Mo.) Post-Dispatch*, December 24, 1883.

Just Moms STL: West Lake Landfill, accessed March 19, 2022. http://www
    .stlradwastelegacy.com/.
Kalin, Berkley. "A History of Ferguson, Missouri, 1855–1918." M.A. thesis,
    Saint Louis University, 1960.
Kay Drey Papers, 1943–2012, Boxes 10 (102716) and 11 (102717). State Histor-
    ical Society of Missouri, Columbia.
Kelly, Cynthia C., ed. *The Manhattan Project: The Birth of the Atomic Bomb
    in the Words of Its Creators, Eyewitnesses, and Historians*. New York:
    Black Dog and Leventhal, 2009.
Khan, Faisal. "Health Advisory: Report of Coldwater Creek Community Ex-
    posures Released." Coldwater Creek Facts, June 25, 2018. http://www
    .coldwatercreekfacts.com/wp-content/uploads/2018/07/Coldwater-Creek
    -Physician-Blast.pdf.
———. "A Legacy of Environmental Health Concerns in St. Louis." Washing-
    ton University in Saint Louis Institute for Public Health, November 30,
    2015. https://publichealth.wustl.edu.
Kirn, Jacob. "Another Lawsuit Accuses Mallinckrodt of Dumping Nuclear
    Wastes, Causing Cancers." *St. Louis (Mo.) Business Journal*, September 8,
    2014.
Koenig, Robert L. "Cleanup Pledge by Bush." *St. Louis (Mo.) Post-Dispatch*,
    February 19, 1989.
Kruse, Peggy. *Old Jamestown across the Ages: Highlights and Stories of Old
    Jamestown, Missouri*. St. Louis County, Mo.: Peace Weavers, 2017.
Krusekopf, H. H., and D. B. Pratapas. *Soil Survey of St. Louis County, Mis-
    souri*. Columbia: University of Missouri, 1919.
Lambert, Thom, and Christopher Boerner. "Environmental Inequity: Eco-
    nomic Causes, Economic Solutions." *Yale Journal in Regulation* 195
    (1997): 196.
Lanquette, William. "Enlisting Einsteins." In *The Manhattan Project: The
    Birth of the Atomic Bomb in the Words of Its Creators, Eyewitnesses, and
    Historians*, edited by Cynthia C. Kelly, 38–41. New York: Black Dog and
    Levanthal, 2009.
"Leadership." St. Louis Lambert International Airport, accessed March 19,
    2022. https://www.flystl.com/about-us/leadership.
Levy, Alexandra. "The Test That Changed the World." *Washington Post*, July 16,
    2017.
Liebow, Edward. "Hanford, Tribal Risks, and Public Health in an Era of
    Forced Federalism." In *Half-Lives & Half-Truths: Confronting the Radio-
    active Legacies of the Cold War*, edited by Barbara Rose Johnston, 145–64.
    Advanced Research Resident Scholar Book. Santa Fe, N.Mex.: School for
    Advanced Research Press, 2007.
Lifton, Robert Jay, and Greg Mitchell. *Hiroshima in America: A Half Cen-
    tury of Denial*. New York: Avon Books, 1995.
"Loyalty Oath of James C. Edwards of Missouri, County of St. Louis," April 1,
    1863. Alamy.com. https://www.alamy.com/949-loyalty-oath-of-james-c
    -edwards-of-missouri-county-of-stlouis-image213440497.html.

Lupkin, Sydney. "Dispute over Missouri Cancer Cluster." ABC News, Febru
    ary 12, 2014. http://abcnews.go.com/Health/dispute-missouri-cancer
    -cluster.story?id=22452352.

Maddox, Robert James. *Weapons for Victory: The Hiroshima Decision Fifty
    Years Later.* Columbia: University of Missouri Press, 1995.

Makhijani, Arjun, and Stephen I. Schwartz. "Victims of the Bomb." In *Atomic
    Audit: The Costs and Consequences of U.S. Nuclear Weapons since 1940,*
    edited by Stephen I. Schwartz, 395–431. Washington, D.C.: Brookings In-
    stitution Press, 1998.

Makhijani, Arjun, Stephen I. Schwartz, and William J. Weida. "Nuclear
    Waste Management and Environmental Remediation." In *Atomic Audit:
    The Costs and Consequences of U.S. Nuclear Weapons since 1940,* edited
    by Stephen I. Schwartz, 353–91. Washington, D.C.: Brookings Institution
    Press, 1998.

"Making the Bomb." Polio Forever (blog), accessed March 19, 2022. https://
    polioforever.wordpress.com/making-the-bomb.

Mallinckrodt, Edward, Jr. Papers, 1798–1981. Missouri State Historical Soci-
    ety, Columbia.

"Mallinckrodt Group, Inc." Company-Histories.com, accessed March 19,
    2022. https://company-histories.com/Mallinckrodt-Group-Inc-Company
    -History.html.

Malloy, Sean. *Atomic Tragedy: Henry L. Stimson and the Decision to Use the
    Bomb against Japan.* Ithaca, N.Y.: Cornell University Press, 2008.

Masco, Joseph. *The Nuclear Borderlands: The Manhattan Project in Post–
    Cold War New Mexico.* Princeton, N.J.: Princeton University Press, 2006.

Mason, Bobbie Ann. "Fallout: Paducah's Secret Nuclear Disaster." *New
    Yorker,* January 10, 2000, 30–36.

"Mayor Arthur F. Bangert." *Municipal Reports Program* of Station KXLW
    (St. Louis, Mo.). Aired Spring 1947.

McCandless, Perry. *A History of Missouri.* Vol. 2, *1820–1860.* Columbia: Uni-
    versity of Missouri Press, 2000.

Merchant, Carolyn, ed. *Major Problems in American Environmental History.*
    Boston: Wadsworth Cengage Learning, 2012.

Middleton, Pat. "Drainage Map of the Mississippi River and Its Tributaries.
    Some Facts and Figures." Ramblin' On, accessed March 19, 2022. www
    .greatriver.com/wordpress/drainage-map-of-the-mississippi-river-and-its
    -tributaries-some-facts-and-figures.

Miscamble, Wilton D. *The Most Controversial Decision: Truman, the Atomic
    Bombs, and Defeat of Japan.* New York: Cambridge University Press, 2011.

"Mississippi River Facts." National Park Service, accessed March 19, 2022.
    https://www.nps.gov/miss/riverfacts.htm.

Missouri Department of Natural Resources Water Protection Program. *Bac-
    teria Total Maximum Daily Load for Coldwater Creek, St. Louis County,
    Missouri, 2014.* Missouri Department of Natural Resources, 2014. https://
    dnr.mo.gov/document-search/coldwater-creek-pathogen-total-maximum
    -daily-load.

Missouri State Gazetteer and Business Directory. St. Louis: Sutherland and
    McEvoy, 1860.
National Service Center for Environmental Publications. *Fact Sheet: Envi-
    ronmental Education Advances Quality Education.* Washington, D.C.:
    United States Environmental Protection Agency, August 1998.
Newman, Richard S. *Love Canal: A Toxic History from Colonial Times to the
    Present.* New York: Oxford University Press, 2016.
Newman, Robert P. *Truman and the Hiroshima Cult.* East Lansing: Michi-
    gan State University Press, 1995.
"The Nobel Prize: Arthur H. Compton Biographical, 1892–1962." The Nobel
    Prize, accessed March 19, 2022. https://www.nobelprize.org/prizes
    /physics/1927/compton/biographical.
North American Association for Environmental Education, accessed March 19,
    2022. https://naaee.org.
"Oak Ridge, TN." Atomic Heritage Foundation, accessed March 19, 2022.
    https://www.atomicheritage.org/location/oak-ridge-tn.
"Office of Environmental Justice in Action." United States Environmental Pro-
    tection Agency, September 2017. https://www.epa.gov/sites/production
    /files/2017-09/documents/epa_office_of_environmental_justice
    _factsheet.pdf.
O'Neill, Kevin. "Building the Bomb." In *Atomic Audit: The Costs and Conse-
    quences of U.S. Nuclear Weapons since 1940,* edited by Steven I. Schwartz,
    33–104. Washington, D.C.: Brookings Institution Press, 1998.
O'Neill, Tim. "After the Great Divorce, How Clayton Became the County
    Seat." *St. Louis (Mo.) Post-Dispatch,* April 19, 2020.
———. "August 22, 1876: How the 'Great Divorce' of St. Louis City and St.
    Louis County Started." *St. Louis (Mo.) Post-Dispatch,* August 22, 2016.
———. "A Look Back: How 35,000 St. Louis Workers Kept Ammo Flowing
    during World War II." *St. Louis (Mo.) Post-Dispatch,* June 27, 2019.
———. "A Look Back: St. Louis Factory Loaded America's Weapons during
    World War II." *St. Louis (Mo.) Post-Dispatch,* June 27, 2010.
———. "Look Back: St. Louis Suburbs Explode after World War II." *St. Louis
    (Mo.) Post-Dispatch,* December 13, 2014.
Paddock Hills—Cincinnati, Ohio. https://paddockhills.org/.
"Paddock Hills, Florissant." *St. Louis (Mo.) Post-Dispatch,* March 6, 1960.
Patrick, Robert. "Bridgeton Landfill Agrees to $16 Million Settlement."
    St. Louis (Mo.) Post-Dispatch, June 30, 2018.
Pitzman's New Atlas of the City and County of Saint Louis, Missouri, 1878.
    Philadelphia: A. B. Holcombe, 1878.
Price, Austin. "Radioactive Waste Could Be Killing Residents in Missouri
    Community, Say Federal Scientists." *Earth Island Journal,* October 30,
    2019. https://earthisland.org/journal/index.php/articles/entry/radioactive
    -waste-could-be-killing-residents-in-missouri.
Primm, James Neal. *Lion of the Valley: St. Louis, Missouri, 1764–1980.* Co-
    lumbia: University of Missouri Press, 1998.

Pritikin, Trisha T. *The Hanford Plaintiffs: Voices from the Fight for Atomic Justice*. Lawrence: University Press of Kansas, 2020.

"Process Used in 1942 to Purify Uranium." *St. Louis (Mo.) Post-Dispatch*, February 12, 1989.

Raasch, Chuck. "Plans to Shift West Lake Cleanup to Corps of Engineers Hits Congressional Roadblock." *St. Louis (Mo.) Post-Dispatch*, July 13, 2016.

"Radioactive Dirt on Highway Came from Chemical Works." *St. Louis (Mo.) Post-Dispatch*, March 2, 1953.

"Radionuclide Basics: Radium." United States Environmental Protection Agency. https://www.epa.gov/radiation/radionuclide-basics.

Ratcliffe, Susan, ed. *Oxford Essential Quotations*. 4th ed. Oxford: Oxford University Press, 2016.

Record of Decision for the North St. Louis County Sites. U.S. Army Corps of Engineers, St. Louis District Office, September 2, 2005.

Rehg, Carol. "Florissant City Council." *St. Louis (Mo.) Globe-Democrat*, October 24, 1979. Kay Drey Mallinckrodt Collection. State Historical Society of Missouri, Columbia.

Rhodes, Richard. *The Making of the Atomic Bomb: 25th Anniversary Edition*. New York: Simon and Schuster Paperbacks, 1986.

———. "Thinking No Pedestrian Thoughts." In *The Manhattan Project: The Birth of the Atomic Bomb in the Words of Its Creators, Eyewitnesses, and Historians*, edited by Cynthia C. Kelly, 19–21. New York: Black Dog and Levanthal, 2009.

Rice, Jack. "The Fordyce Mansion Makes Its Last Stand." *St. Louis (Mo.) Post-Dispatch*, July 25, 1963.

Rome, Adam. *The Bulldozer in the Countryside*. Cambridge: University of Cambridge Press, 2001.

Rotter, Andrew J. *Hiroshima: The World's Bomb*. Oxford: Oxford University Press, 2008.

Rust, Daniel L. *The Aerial Crossroads of America: St. Louis's Lambert Airport*. St. Louis: Missouri History Museum Press, 2016.

Salter, Jim. "West Lake Landfill Cleanup Slows after More Nuclear Waste Found." *St. Louis (Mo.) Post-Dispatch*, March 19, 2022.

Sarathy, Brinda, Vivien Hamilton, and Janet Farrell Brodie. *Inevitably Toxic: Historical Perspectives on Contamination, Exposure, and Expertise*. Pittsburgh, Pa.: University of Pittsburgh Press, 2018.

Satterfield, Theresa, and Joshua Levin. "From Cold War Complex to Nature Preserve: Diagnosing the Breakdown of a Multi-Stakeholder Decision Process and Its Consequences for Rocky Flats." In *Half Lives & Half-Truths: Confronting the Radioactive Legacies of the Cold War*, edited by Barbara Rose Johnston, 165–91. Santa Fe, N.Mex.: School for Advanced Research Press, 2007.

Schanzenbach, Diane Whitmore. "Cancer Study Surveyed the Wrong People." Northwestern Now, April 5, 2013. https://news.northwestern.edu/stories/2013/04/opinion-st.-louis-beacon-schanzenbach/.

Schanzenbach, Diane Whitmore, and Jenell Rodden Wright. "Commentary: State Department of Health Cancer Study Surveyed the Wrong People." St. Louis Public Radio, March 26, 2013. https://news.stlpublicradio.org /health-science-environment/2013-03-26/commentary-state-department -of-health-cancer-study-surveyed-the-wrong-people.

Schein, Edgar H. *Organizational Cultures and Leadership.* 2nd ed. San Francisco: Jossey-Bass, 2004.

Schneider, Keith. "Mountain of Nuclear Waste Splits St. Louis and Suburbs." *New York Times*, March 24, 1990.

Schroeder, Walter. "Environmental Setting of the St. Louis Region." In *Common Fields: An Environmental History of St. Louis*, edited by Andrew Hurley, 13–37. St. Louis: Missouri Historical Society Press, 1997.

Schuessler, Ryan. "Missouri Home Contaminated by Wartime Radioactive Waste, Lawsuit Says." *The Guardian*, November 16, 2016.

Schwartz, Stephen I. "Congressional Oversight of the Bomb." In *Atomic Audit: The Costs and Consequences of U.S. Nuclear Weapons since 1940*, edited by Stephen I. Schwartz, 485–517. Washington, D.C.: Brookings Institution Press, 1998.

"7 More Nuclear Waste 'Hot Spots' Found in St. Louis Suburb." CBS News.com, December 22, 2015. https://www.cbsnews.com/news/seven-more -nuclear-waste-hot-spots-found-in-north-st-louis-county-missouri -suburb/.

"7 Things You Should Know About Poverty and Housing." Habitat for Humanity, accessed March 20, 2022. https://www.habitat.org/stories/7 -things-you-should-know-about-poverty-and-housing.

Siouxland News at Sunrise. KMEG (CBS, Sioux City, Iowa), December 23, 2015.

Skates, John Ray. *The Invasion of Japan: Alternative to the Bomb*. Columbia: University of South Carolina Press, 1994.

Smith, Irene Sanford. *Ferguson: A City and Its People*. Ferguson, Mo.: Ferguson Historical Society, 1976.

Smith, Kathleen Logan. "Coalition Report." *Healthy Planet*, November 1, 2013. http://thehealthyplanet.com/2013/11/coalition-report-26/.

"State Cancer Registries: Status of Authorizing Legislation and Enabling Regulations—United States, October 1993." Centers for Disease Control and Prevention. https://www.cdc.gov/mmwr/preview/mmwrhtml /00023762.htm.

Steingraber, Sandra. *Living Downstream: An Ecologist's Personal Investigation of Cancer and the Environment*. Philadelphia: Da Capo, 2010.

Stimson, Henry L. "The Decision to Use the Atomic Bomb." *Harper's Magazine*, February 1947, 97–107.

———. "The Decision to Use the Atomic Bomb." In *The Manhattan Project: The Birth of the Atomic Bomb in the Words of Its Creators, Eyewitnesses, and Historians*, edited by Cynthia C. Kelly, 383–88. New York: Black Dog and Levanthal, 2009.

"St. Louis Airport Site: Site History." U.S. Army Corps of Engineers, St. Louis
District Website, accessed March 19, 2022. https://www.mvs.usace.army
.mil/Missions/FUSRAP/SLAPS.

"St. Louis Area Residents Sue over Radioactive Material Storage Sites." *St.
Louis (Mo.) Post-Dispatch*, February 21, 2018.

St. Louis County Postwar Subdivisions Study, 2003. Missouri State Parks
website, posted by Missouri State Preservation Office, January 17, 2018.
https://mostateparks.com/sites/mostateparks/files/STLC%20Postwar
%20Subdiv%20Survey.pdf.

"St. Louis Sites Fact Sheet: Frequently Asked Questions about FUSRAP."
U.S. Army Corps of Engineers, St. Louis District Website, June 30, 2020.
https://www.mvs.usace.army.mil/Portals/54/docs/fusrap/factsheets
/Virtual_FUSRAP_Open_House/FAQ_Fact_Sheet.pdf.

St. Louis Site Remediation Task Force Report. St. Louis, Mo.: U.S. Army
Corps of Engineers, St. Louis District, 1996.

Studies of Aquatic Life. 2003 Corps Report. Vol. 1, *Feasibility Study for the St.
Louis North County Site*. Army Corps of Engineers, May 1, 2003. https://
www.mvs.usace.army.mil/Portals/54/docs/fusrap/docs/FSNCounty_1.pdf.

Suburban Stats: Population Information and Statistics from Every City, State
and County in the US, 2019. suburbanstats.org.

Swain, Frank. "The Congo's Role in Creating the Bombs Dropped on Hiroshima and Nagasaki Was Kept Secret for Decades but the Legacy of Its Involvement Is Still Being Felt Today." BBC, August 3, 2020. https://www
.bbc.com/future/article/20200803-the-forgotten-mine-that-built-the
-atomic-bomb.

Tarr, Joel A. *The Search for the Ultimate Sink: Urban Pollution in Historical
Perspective*. Akron, Ohio: University of Akron Press, 1996.

Tarr, Joel A., and Carl Zimring. "The Struggle for Smoke Control in St. Louis."
In *Common Fields: An Environmental History of St. Louis*, edited by Andrew Hurley, 199–220. St. Louis: Missouri Historical Society Press, 1997.

Theising, Andrew J., ed. *In the Walnut Grove: A Consideration of the People
Enslaved in and around Florissant, Missouri*. Florissant, Mo.: Florissant
Valley Historical Society, 2020.

Thomas, Takesha. "St. Louis Man's Alleged Radiation Exposure Spurs
Wrongful Death Lawsuit." *St. Louis (Mo.) Record*, July 23, 2018.

Tighe, Theresa. "Carolyn Bower Remembers Lou Rose." The First Secret City,
October 15, 2015. https://firstsecretcity.com/2015/10/15/carolyn-bower
-remembers-lou-rose.

"Tom Green Interview-Excerpts, 1979." St. Louis Area Downwinders: Coldwater Creek Facts. http://www.coldwatercreekfacts.com/media
/downwinder.php.

"Transfer of Radioactive Dirt Leaves Berkeley with Doubts." *St. Louis (Mo.)
Globe-Democrat*, September 19, 1978.

Truman, Harry S. "The Battle of the Laboratories." In *The Manhattan Project:
The Birth of the Atomic Bomb in the Words of Its Creators, Eyewitnesses,*

and Historians, edited by Cynthia C. Kelly, 339–42. New York: Black Dog and Levanthal, 2009.

Tuft, Carolyn. "Airport Seeks All of Kinloch." *St. Louis (Mo.) Post-Dispatch*, May 2, 1995.

United States v. State of Missouri, 388 F. Supp. 1058 (E.D. Mo 1975). Justia, 2022. https://law.justia.com/cases/federal/district-courts/FSupp/388 /1058/2313211/.

"Uranium and Depleted Uranium." World Nuclear Association, updated November 2020. https://world-nuclear.org/information-library/nuclear -fuel-cycle/uranium-resources/uranium-and-depleted-uranium.

"Uranium Mining." Atomic Heritage Foundation, July 30, 2018. https://www .atomicheritage.org/history/uraniummining.

U.S. Army Corps of Engineers, St. Louis District FUSRAP. www.mvs.usace .army.mil.

U.S. Department of Energy. The Manhattan Project: An Interactive History. https://www.osti.gov/opennet/manhattan-project-history/.

U.S. Department of Health and Human Services Public Health Service Agency for Toxic Substances and Disease Registry, Division of Community Health Investigations. *Evaluation of Community Exposures Related to Coldwater Creek St. Louis Airport / Hazelwood Interim Storage Site, North St. Louis County, Missouri*. Atlanta, April 30, 2019.

U.S. Environmental Protection Agency. "National Environmental Education Act," November 16, 1990. United States Environmental Protection Agency. https://www.epa.gov/sites/default/files/documents/neea.pdf.

"A Very Sobering Event." In *The Manhattan Project: The Birth of the Atomic Bomb in the Words of Its Creators, Eyewitnesses, and Historians*, edited by Cynthia C. Kelly, 329–30. New York: Black Dog and Levanthal, 2009.

Vierzba, Edmund A., to Arthur Whitman, August 20, 1985. Kay Drey Mallinckrodt Collection. State Historical Society of Missouri, Columbia.

"Village of Hazelwood, St. Louis County, Mo." St. Louis County Library Headquarters, Ladue, Mo.

Wainstock, Dennis D. *The Decision to Drop the Atomic Bomb*. Westport, Conn.: Praeger, 1996.

Walker, J. Samuel. "Historiographical Essay: Recent Literature on Truman's Atomic Bomb Decision; A Search for Middle Ground." *Diplomatic History*, 29, no. 2 (2005): 311–34.

——— . *Permissible Dose: A History of Radiation Protection in the Twentieth Century*. Berkeley: University of California Press, 2000.

——— . *Prompt and Utter Destruction: Truman and the Use of Atomic Bombs against Japan*. Chapel Hill: University of North Carolina Press, 2004.

——— . *A Short History of Nuclear Regulation, 1946–1999*. North Bethesda, Md.: United States Nuclear Regulatory Commission, 2000.

Wammack, Mary D. "Atomic Governance: Militarism, Secrecy, and Science in Post-war America, 1945–1958." PhD diss., University of Nevada, Las Vegas, 2010.

Wanko, Andrew. "The Great Divorce." Missouri Historical Society, https://
mohistory.org.

Weller, George. "The Atomic Bomb's Peculiar 'Disease.'" In *The Manhattan
Project: The Birth of the Atomic Bomb in the Words of Its Creators, Eye-
witnesses, and Historians*, edited by Cynthia C. Kelly, 356–59. New York:
Black Dog and Levanthal, 2009.

"West Lake Landfill." Missouri Department of Natural Resources, accessed
March 20, 2022. https://dnr.mo.gov/waste-recycling/sites-regulated
-facilities/federal/west-lake-landfill.

"What Is Environmental Education?" United States Environmental Protection Agency, accessed March 20, 2022. https://www.epa.gov/education
/what-environmental-education.

Williams, Susan. *Spies in the Congo: The Race for Ore That Built the Atomic
Bomb*. London: Hurst, 2016.

Williams, Terry Tempest. *Refuge: An Unnatural History of Family and Place*.
New York: Vintage Books, 2018.

Wohlforth, Jenny. "What Lies beneath the Fernald Preserve." *Cincinnati
Magazine*, June 7, 2010.

World Population Review, accessed March 20, 2022. https://worldpopulation
review.com/us-cities.

"World War II and St. Louis." AboutStLouis.com, accessed March 20, 2022.
https://aboutstlouis.com/local/history/world-war-2-st-louis.

Wright, Janell Rodden. Interview with author. February 5, 2020.

Wright, John A., Sr. *Kinloch: Missouri's First Black City*. Chicago: Arcadia,
2000.

———. *St. Louis: Disappearing Black Communities*. Chicago: Arcadia, 2004.

Yun, Shumei, Chester Lee Schmaltz, Sherri Homan, Philomina Gwanfogbe,
Sifan Liu, Jonathan Garoutte, and Noaman Kayani, et al. *Analysis of Cancer Incidence Data in Coldwater Creek Area, Missouri, 1996–2004*. Missouri Department of Health and Human Services. health.mo.gov.

Zaretsky, Natasha. *Radiation Nation: Three Mile Island and the Political
Transformation of the '70s*. New York: Columbia University Press, 2018.

Zeiler, Thomas W. *Unconditional Defeat: Japan, America, and the End of
World War II*. Wilmington, Del.: Rowman and Littlefield, 2003.

Zhang, Sarah. "The Secret 1949 Radiation Experiment That Contaminated
Washington." *Spokane (Wash.) Chronicle*, June 3, 2015.

INDEX

Printed in the United States
by Baker & Taylor Publisher Services